Vortex Publishing LLC.
4101 Tates Creek Centre Dr
Suite 150- PMB 286
Lexington, KY 40517

www.vortextheory.com

© Copyright 2019 Vortex Publishing

All rights reserved. No part of this book may be reproduced or transmitted in any form or by any means, electronic or mechanical, including photocopying, recording or by any information storage and retrievable system without the prior written permission by the Publisher. For permission requests, contact the publisher.

Printed in the United States of America

1 2 3 4 5 6 7 8 9 10

Library of Congress Control Number: 2019953616

ISBN   978-1-7332996-8-8
eISBN 978-1-7332996-9-5

Editor's note: All drawings in this book are original illustrations made by Dr. Moon. They are kept as they are to maintain the integrity of his work.

TABLE OF CONTENTS

Publisher's Note III
Author's note IV
Forward V

PART I
I WAS TAUGHT THE SKILLS I NEEDED, AND I WAS TRAINED TO WIN

Chapter 1: It Was 1967→The Summer of Love in San Francisco… 1
Chapter 2: The Underwater Demolition Teams… 4
Chapter 3: Boolean Algebra: The Science of Logic… 7

PART II
THE QUEST BEGINS…

Chapter 4: The Near Death Experience in Readers Digest Magazine 9
Chapter 5: Where Is the Kingdom of God Located? 14
Chapter 6: The Search for the Kingdom of God Begins … 19
Chapter 7: Higher Dimensional Space 22
Chapter 8: The Greatest Story That's Never Been Told... 27
Chapter 9: There Is a Problem With the Scientific Vision of the Universe! 30
Chapter 10: Hard Times… 32

PART III
EINSTEIN'S MISTAKE…

Chapter 11: I Find the Mistake in the Theory of Relativity 37
Chapter 12: The Curious Relationship Between Time and Motion 42
Chapter 13: The Michelson Morley Experiment 46
Chapter 14: The International Association for New Science.. 52

PART IV
POWERFUL ALLIES COME TO MY AID…

Chapter 15: A Russian Scientist Was Sent to the Conference in Fort Collins! 55
Chapter 16: Fate Brings Me Another Powerful New Ally…Her Name Is Jeannie! 60
Chapter 17: A Great Russian Scientist Becomes My Ally…and Mentor! 64

PART V
I DEFEND THE THESIS

Chapter 18: Fabian and I Go to Russia…... 69
Chapter 19: We Increase the Speed of Light … 71
Chapter 20: I Defend the Thesis 74

PART VI
THE LEGACY OF THE VORTEX THEORY OF ATOMIC PARTICLES

Chapter 21: The Legacy of the Vortex Theory of Atomic Particles 78
Chapter 22: Legacy Continued: Quantum Entangled Technology
          For Cell Phones Is Now Possible 82
Chapter 23: Legacy Continued: Anti-gravity Engineering Is Now Possible …!!! 86
Chapter 24: I Meet Stan Clifford. 91
Chapter 25: Flashback to the Past…I Discover the Secret of How to Interpret Jesus' Words,
          Statements, and Parables From His Own Point of View, as if He Were Standing
          Right in Front of Us Explaining Them to Us!.. 98
Chapter 26: Working 24/7 for a Year, I Finally Make the Analysis!... 106

REFERENCES & INTERNATIONAL CONFERENCES 109
RUSSIAN SCIENTIFIC JOURNALS 114

EPILOGUE
Synchronicities That Lead Russell to the Truth.. 115

Publisher's Note

I am Stan Clifford, CEO of Vortex Publishing. I am also CEO and partner of Dr. Russell Moon in V.T.I., LLC which is a company we formed in 2015 to bring to the world the fruits of Dr. Moon's most important discoveries.

When starting this, we decided to form Vortex Publishing so that Dr. Moon could re-publish and update his first book, "The End of the Concept of Time" and write other books to further explain the Vortex Theory and all of the discoveries he has uncovered by using the concepts of the Vortex Theory. This series is now a six part series.

This book, "Part V, How It All Began" is a Book I asked Dr. Moon to write because the story of who he was is an amazing story in itself. How could the most unconventional physicist of all time perhaps become the most famous and successful physicist of all time? If you glance quickly at pages 74 through 77 and look at the sheer volume of this man's discoveries, it will make your jaw drop. Three of the most significant discoveries of the Vortex Theory aren't even listed on these pages …, the first being that this theory has discovered not only that higher dimensional space exists but how it exists. The second major discovery is that the structure of the atom is totally different than the world of science thinks. Protons and electrons are not solid particles, but three dimensional holes that space flows through at the speed of light. In one direction the space is in 3 dimensional space and in the other direction it is in higher dimensional space. The third discovery that is not listed in this book is the fact that every electron in the universe possesses an anti-gravity component and yes, this component can be harnessed to create a whole new totally green power source that can save the world from climate change among other things. These three discoveries, in my opinion, are the greatest scientific discoveries of all time!!! These discoveries are the primary reason Dr. Russell Moon was awarded the equivalent of a PhD. in Nuclear Physics by the government of Russia. This book also explains the rigors of approval Dr. Moon experienced in Russia to achieve this award.

For me to meet Dr. Moon was just another of the weird synchronicities of his life that have helped him achieve his greatness. Reading the story of the near-death experience, randomly meeting his wife Jeannie and the top physicists of Russia were just three other synchronicities that helped Dr. Moon achieve what he has done. Dr. Moon possesses a very special intellect. His brain can think of concepts most men cannot even fathom. What kind of person spends his whole career trying to prove higher dimensional space exists? In this case, that discovery has and will change all physical science forever. This is his story.

Stan Clifford

CEO, V.T.I., LLC

Vortex Publishing, LLC

AUTHOR'S NOTE:

No knowledge of science or mathematics is needed for the reader of this book. It tells the story of how a mysterious, rogue physicist came out of nowhere and changed the world of science forever. Because some parts of the story were already told in the very first book, *The End of the Concept of "Time,"* and there was no easy way to rewrite them, they were paraphrased in this one.

The purpose of this book is to explain how and why I undertook a life-long quest to discover the knowledge sought from the beginning of recorded history by all the world's greatest philosophers, sages, and scientists…and how I succeeded!

Because all of the scientific discoveries are in the first three books of this series, they do not need to be repeated again here. But instead, just the story of how it all happened is being recounted for all to learn.

In honor of two of Russia's greatest scientists: Dr., Prof. Konstantin Gridnev; and Dr., Prof. Victor V. Vasiliev, I have given the vortices the name "Konsiliev Vortices".

Russell Moon

Forward

One evening I was relaxing while listening to some 60's folk music by the Kingston Trio, and Bob Dylan, when I got a phone call from my good friend Stan Clifford. Stan, who had originally pushed me to write the last two books on the new and revolutionary scientific knowledge that has come to be called, *"The End of the Concept of Time – Trilogy;"* he went on to say he was calling to convince me to write one more book.

He told me that during his recent talks with various physics professors, the question that always seemed to come up again and again was how an individual without any advanced mathematical training beyond Differential Equations had nevertheless succeeded in discovering the key, the secret, to explaining <u>all</u> of the great mysteries about the construction of the universe? Mysteries that some of the greatest physicists in the world had spent their entire lives working on, yet could not explain: men like Albert Einstein, Wolfgang Pauli, Werner Heisenberg, Stephen Hawking, Arthur Eddington, Linus Pauling, Carl Jung, Paul Dirac, Arnold Sommerfeld, Max Planck, Edwin Hubble, Niels Bohr, Richard Feynman, and many, many more. And again: the question that kept arising was how come not one of these famous scientists, many of whom were Nobel Prize winners, could explain what Dr. Moon has seemingly done so simply and without any effort whatsoever?

He asked me, "Didn't Book 1, *The End of the Concept of Time* adequately explain everything about how it all happened?" He emphatically stated, "No! Absolutely not! It explained some, but nobody knows what happened before or after your discoveries. Nobody knows how a loser ["No offense!"] who barely graduated from High School was somehow miraculously transformed into a winner, later graduating from College with straight "A's" and honors. Nor does anyone know about the almost mystical chain of events that happened after the first book was written; about how former 'hard core' Atheistic Russian scientists came to your aid and threw their support behind a scientific thesis based upon the words of Jesus in the New Testament! How a work of *science* transformed them into Christians! It is unprecedented! There is nothing like it!"

He went on to say, "Even more shocking is the fact that the Russian Ministry of Education then awarded you a PhD for this thesis [in Russia, all higher academic degrees are awarded by the government]. A thesis that contradicted everything the former Communist Government once stood for! Not only was your thesis revolutionary, it is even more revolutionary to learn it was published by a government that once shuttered churches, arrested priests, and was against everything Christ stood for! A lot of people want to know even more about *these* events than your scientific discoveries!"

He then went on to tell me, "…that whether you like it or not, you are not only becoming a famous figure in physics today, but eventually, you will be a famous historical one as well! After your death, many future generations will want to know your story"! [Oh Boy!]

Stan is not only an excellent CEO for his successful company DecoArt in Stanford Kentucky, but also an award-winning entrepreneur from his alma mater Clemson University and a very influential speaker [and talker, I know!]. From past experience, I have learned that his wise counsel and heartfelt guidance is worth listening to. So, I told him I would think about it.

Later, I got to thinking about what he said and realized that in fact it was true: just how did it happen that a nobody, a loser, came out of nowhere to successfully challenge and change the 20th Century world of science? Although I now accept the responsibility bestowed upon me, when it originally fell to me, I did not want any part of this avant-garde role. I merely wanted to find the scientific and religious answers I had dedicated my life to finding. Answers I had to know!

You see, I was blessed [or maybe cursed?] by a haunting question that continually troubled me day and night my entire adult life and would not go away, no matter how hard I tried to ignore it... *I had to know the truth: were we indeed just a bunch of intelligent animals, or was there really a purpose and reason for being alive?*

So, which was true: was it the cold blooded Atheistic vision I had grown up with; the one that says we are just some sort of biological accident – an infestation of intelligent microbes crammed together, living upon a little grain of sand in the middle of a vast and frighteningly large universe? Or was it the "religious" viewpoint: the belief that there is indeed a purpose, a reason for being alive; that we possess a second invisible part of us called the soul; that there is a supreme power in the universe, a powerful consciousness, a God who created everything…and there is a Kingdom of souls where our soul goes to at death?

I also thought about how I used revolutionary methods to break with conformity, to force myself to think differently from everyone else and in the process, make all of these incredible scientific discoveries. If these unconventional methods were finally revealed, it might help and motivate other future seekers in their quests too. It might also re-invigorate those who have given up; who have looked at the magnitude of the problem facing them and reluctantly backed away, considering themselves too inadequate or ignorant to confront the enormity of the task of being a Martin Luther and going against the beliefs of an entire culture! But now, hearing my story might give them new hope and a new reason to carry on, to succeed: to WIN!!!

And finally, and maybe the most important of all, the unexplainable, mystical help that came to me in times of despair; or when I wanted to quit, give up, and escape from this unrelenting, unstoppable quest that consumed my entire adult life. This supernatural help would come to me in times of need, almost as if I were being watched over and cared for by some unseen super powerful force!

So, after much consideration, I came to the conclusion that Stan Clifford's intuitive feeling about the need to tell this story was correct. And so…I have decided to tell the story about how a simple person of no-consequence [or so I have been told by invidious academics] discovered the revolutionary scientific knowledge about the construction of the universe. Knowledge whose advanced future technology will one day not only change the lives of all the peoples on the earth, but will eventually allow us to leave the safety of the earth in anti-gravity vehicles and travel to the other planets of this solar system, and then finally, to the stars beyond.

PART I
I WAS TAUGHT THE SKILLS I NEEDED, AND I WAS TRAINED TO WIN…

Chapter 1
It Was 1967→The Summer of Love in San Francisco…

Most people's lives begin at birth, but mine began one day after graduating from High School when I went to a military recruiter's office in the town of Costa Mesa California and joined the Navy.

Up until that time, it is painful to report that I had been a typical lazy teenager with no goals, ambition, or purpose in life. I only joined the Navy because my next door neighbor who had been a Major in General Patton's Third Army told me about an excellent program the Navy was offering to train sailors to be Electronics Technicians; giving them a good marketable skill for future employment when they got out of the service. Because he was earnest in believing it was a good program and talked about the added incentive of traveling to exotic ports in foreign countries and seeing the world, it sounded like fun. Also, since I knew I did not want to go to college, I though it seemed like a good deal. A few days later, I realized I made a serious mistake!

Joining the Navy was not the idyllic vacation I believed it was going to be. <u>For the first time in my life, I was taught strict discipline</u>. Up at 4:30 in the morning to get to the chow hall by 5:00. Then training and more training all day long about what sailors need to know about ships, navigation, and the tying of knots! Then there was exercising, marching, cleaning barracks, and every other skill a good military man needed to know.

<u>Like most veterans I have talked to since, I would never want to go through "boot-camp" again, but I would never trade all of this training and self-discipline for anything else</u>. It is true; the United States Military takes boys and turns them into men. And so, it was with me. Unfortunately, after boot-camp I was sent to Electronics Tech School in San Francisco California.

I say unfortunately because it was 1967, "The Summer of Love" in the United States, and San Francisco was ground zero for all of it. Protests, riots, pot smoking and hippies were the rules of the day. Wine was cheap and pot was cheaper. I did not know anyone who did not try smoking pot or drinking wine [especially, "Boone's Farm Strawberry Hill, and Thunderbird"]. It was the thing to do, and San Francisco was the place to do it.

Everyone was there: Janis Joplin, Big Brother and the Holding Company, Jimmy Hendrix, the Doors, and Jefferson Airplane just to name a few. They played at the Forum at night, and when they weren't playing they raised hell. I, like the rest of the students at the Electronics school located on Treasure Island in the middle of the San Francisco Bay, learned both electronics and how to party. It was a hell of a time.

Everything centered around the Golden Gate Park and Haight Ashbury. Me and two buddies, while studying at night, used to listen to the radio, and if we heard about a riot, immediately dropped the books, got onto a bus, and headed to Haight Ashbury. It was a lot of fun; I never knew why Tear-gas was called "Tear"-gas until then; but sometimes too much fun means less time to study. My grades suffered and I got set back several times and had to take some of the courses over again.

Some of the foolishness I experienced stands out above the rest. Such as the time when about 50 of us tried to get medical discharges by convincing the base psychiatrist we were all crazy. It started like this: a student at the school – call him Gomer – while cleaning out his locker, told us he was getting a medical discharge out of the service. When asked how he did it, he went on to say that all you had to do was to go to the base dispensary, see the psychiatrist and tell him you were crazy. So, the next morning about 50 of us lined up outside the dispensary all wanting to see the psychiatrist. Unfortunately, I was nearly last in line, so by the time I got to see this [by then] distraught doctor in his office, he was already wise to the ruse and completely out of patience.

When it was my turn to see him, before I even had time to close the door as I entered into his office or say anything he said, "OK fella, just what the hell is your story?" Taken by surprise I told him the "dim-witted" answer I had worked out while standing in line by saying, "…anyone who joins the military in the time of war is crazy." To which he said, "I joined the God Damm Military in the time of war! Am I crazy too?" To which I should not have said anything but blurted out without thinking, "Yes! You too must be as crazy as a Bat in a Belfry!" To which he replied yelling, "GET OUT! GET THE HELL OUT OF HERE!!!" [Of which I did very fast!] Well so much for being crazy. All 50 of us ended up doing extra duty as a reward for just trying to help rid the Navy of a few undesirables like ourselves: or so we remorsefully lamented while drinking a few beers at the Enlisted Men's Club. [Later, we all found out from our Chief Petty Officer that "Gomer" had lied: he did not say he was crazy, but instead a homosexual, and that's why they sent him home so fast!]

<u>Nevertheless, this was a valuable learning experience, for even though it did not turn out well, it was the first time in my life I had ever challenged authority</u>. Another valuable lesson was learned during the long tumultuous months of the "Summer of Love" in 1967.

The summer of 1967 was filled with wild antics that continuously took place in the vicinity of Haight Street and the Golden Gate Park. It started with the tragedy of the invasion of over one hundred thousand "long hairs" from all over the country. Vagrants calling themselves hippies had come to San Francisco with no money or any place to stay and ended up living on the Fillmore District's back streets and alley-ways. Next there were the constant daily and sometimes violent protests against the Viet Nam War. And to top it all off, the ironic protests of the original hippies against the newcomer hippies.

These original hippies staged the mock "Death of the Hippy Funeral," by marching down Haight street with an open casket to express their disapproval against the thousands of want-to-be hippies crowding into Frisco and trying to copy them by wearing love beads, sandals, and colorful Moroccan shirts. Many were lost souls, good kids who had thoughtlessly left home, parents, and school after listening to Timothy Leary's insane message of …"turn on; tune in; and drop out;" and his even more insane insistence that taking LSD is good for you! Good kids not realizing the self-destructive implications of their impulsive thoughtless choice of dropping out to what, and to go where? [To go to San Francisco to beg for spare change, and sleep on the streets?]

It was both shocking and depressing at the same time to see my generation trying to be different from all previous generations. Yet ending up being lost souls going nowhere! What a tragedy!

Luckily, being in the Navy, I could look into this insane kaleidoscope of directionless "Happenings" without falling in and becoming trapped myself! I could visit the problem without becoming part of the problem by going back to the base at night and sleeping in my bunk. In doing so, I could walk away from it all without being changed by it.

However, I did gain something special from all these experiences I now consider to be priceless: I learned from the example of the boldness of the many protestors I witnessed that sometimes, just sometimes standing tall upon a soapbox with a fist in the air and being defiant is OK. That the negative opinions of those who initially think you are wrong and are booing you or throwing garbage at you aren't that important anyway. It is what *you* think that eventually counts in the end.

<u>It was a valuable lesson I was able to apply later in life, when I had to boldly, and courageously stand up myself before groups of unruly scientists and tell them the cold hard facts that what they and their colleagues are teaching to millions of students all over the world about the construction of the universe is totally and completely wrong!</u>

Yes, I realize now that I learned a lot from attending Electronics Tech School in San Francisco!

Chapter 2
The Underwater Demolition Teams…

Sometimes the seemingly mundane things we do today only reveal their importance later on in life. After graduating from Electronics Tech School, I was sent to the Navy's Anti-submarine Warfare Airport in Brunswick Maine. I worked in the Control Tower repairing the electronics the Air Traffic Controllers used for talking to the constantly landing and taking off P-3 Squadrons: the anti-submarine warfare planes patrolling the North Atlantic.

Because I was mostly on call nearly all of the day, with my main job being ready to make instant emergency repairs when needed, I only did light maintenance and repairs on the 24 hour tape recorders recording the conversations between pilots and the air traffic controllers and nothing else. So, to occupy my time and keep from dying of boredom in my repair laboratory, most of the day, I alternated between practicing juggling tennis balls and exercising doing push-ups and sit-ups. I got quite good at push-ups and sit-ups, doing 200 at a time of the first, and 500 at a time of the second. I eventually added running in place to the routine along with doing chin-ups on a bar I placed between two large electronics cabinets. When the semi-annual "Iron Man" contest was held on the base, I took first place two times in a row, beating out over 3,500 men. I not only got 3 days special leave for each victory, but gained the attention of the Navy's Atlantic Underwater Demolition Teams located in Little Creek Virginia.

One of their lieutenants personally called me to ask if I was interested in joining the UDT, and when I said "sure," the Lieutenant said he was going to fly up to Maine to administer the physical fitness test. He said on the phone the physical fitness test was "easy": one only had to do 20 push-ups, 30 sit-ups, and 10 chin-ups in 90 seconds [for me, nothing to it!]; then you had to run a mile in 6 ½ minutes [again nothing to it – I could run a mile in 5 minutes]; and finally swim 300 yards also in 6 ½ minutes using the underwater breaststroke, sidestroke, and backstroke; then dive down to the bottom of the pool and hold your breath for 15 seconds. Since I had grown up on the beaches of Southern California, I thought the swimming test was going to be easy and the rest was going to be a breeze. So much for thinking!

The lieutenant couldn't make it [or so he said], so instead, they sent a Navy SEAL in his place: a Master Chief Petty Officer with a bad attitude at best. I did the push-ups, sit-ups, and chin-ups in 90 seconds with ease. But then the Chief said he did not like my form on the chins, so to do it all over again. Then after doing it all over again, he said he still did not like my form so to do it again. Then again and again and again and again until I was nearly exhausted, barely able to finish just a couple of chins at the last go of it; and then, [he was sizing me up] just before I was going to tell him to go to hell, he said, "Let's try the mile?"

A course had been laid out on the dirt road behind the Airport runway and we took off: me running while he drove a jeep *directly* behind me. If I tripped and fell, I would have gotten run over! But after I passed the mile marker in just under 5 minutes, he told me, "I didn't like your form! Do it again!!!" And so, I did it again and again and again and again.

And just like before, when I was just about ready to tell him to go to hell he said, "Let's try the pool!"

And just like before, I did the swimming test again and again and again. The last time I got out of the pool, I just walked right on past him without even looking at him or saying anything and was headed directly to the locker room when he yelled back at me saying, "Hey, you passed!" [I did not know then that he was not only testing athletic ability, but also the ability to persevere and endure the severe physical stress and emotional rigors of the Navy's Frogman School.]

Although I didn't want to admit it in front of the Chief… this was great! I wanted to be a Navy Frogman. It was the fun I was looking for when I originally joined the Navy. The thought of swimming to beaches and blowing up underwater obstacles before invasions seemed like great fun. However, I had a serious problem I never told anybody about, I had very poor eyesight. In fact, I was almost blind in one eye [20/400 in my right eye where 20/20 is deemed normal]! At night, without my glasses, I could not see with my right eye, everything was a blur. Nevertheless, I faked the eye test during the base doctor's physical examination at the Brunswick Naval Air Station by previously memorizing all the eye charts shown to me by a friend of mine; and I easily passed the diving bell test in Portsmouth Virginia, and the psychiatric evaluation; but it was only later that I realized the hard truth about just how dangerous my handicap was.

I had several sets of hard contact lenses hidden in a secret pocket sewn inside my swimsuit, but after losing one out of my eye while swimming in the open ocean, I realized that I could not even see the barge I was swimming to. Equally difficult was trying to replace the hard contact lens while swimming in the ocean with waves constantly hitting your face. [Soft contact lenses were just being developed by Bausch and Lomb, but not yet on the market.] So sadly, I was finally confronted with the lie I was living and knew then the awful truth that I could not continue.

Eventually I had to use the excuse of having a damaged Achilles tendon to say I was quitting. It was the worst day of my life. But it was only years later that I realized it was also the best day of my life. A friend who also had to drop out because of a broken bone in his foot was sitting on his bunk in the barracks, consoling himself by reading a book called "*There is a River*," the story of Edgar Cayce, America's famous sleeping prophet.

I asked this friend who had studied physics in college what a scientist was doing reading a book on parapsychology. He told me something like this, "It might be deemed parapsychology today, but the subject has not been adequately studied and just might be the physics of the future?" I was so impressed by his answer, I ended up reading the book after he finished. This book became very important to me later on in life because it bridged the gap between science and religion; opening my mind, and for the first time in my life, making me question just how true were the Atheistic beliefs I was brought up with?

The Navy's UDT training was also invaluable later on in life because it was the only training in the world that taught you how to win → to succeed no matter what the odds were that faced you. It taught you how to persevere, to endure, to never give up. You are divided up into 7 man teams. Each team carries a large 7 man rubber raft on top of their helmeted heads, and you run everywhere on the base with it [there is no walking!]. The teams are then pitted against each other from morning until night in a host of contests. If you win a contest your team gets to rest while everyone else has to do the race or exercise over and over again. So real quick, you learn that you have to give everything you have in one gigantic burst of energy to win, to succeed the first time, every time. That is the only way you can make it through the day. If you fail early in the day, the repetition of doing the exercise over and over again makes you and your whole team become tired early, and then you will continue to fail the rest of the day because everyone else is always fresher than you are.

You are also psychologically challenged all day long. Instructors continually scream directly in your face; call you every name in the book [and some that aren't in the book!]; and hit you on your helmet indiscriminately with their fists. You learn to laugh at this psychological warfare and not let it affect you in the least; and never, never even consider giving them the satisfaction of making you quit: [some did!].

But when I realized I had to quit too because I knew if I lost a contact lens in the water, I could never make the open ocean swim of 7 miles I would be expected to be able to do at the Roosevelt Roads Naval base in Puerto Rico, it was a bitter, bitter lesson. [This all happened before Lasik Eye Surgery was invented many years later to restore 20/20 eyesight.]

Nevertheless, this bitter lesson was used to my advantage later on in life while struggling year after year to somehow mathematically prove the shocking truth I had discovered about "time" not existing as a fundamental principle of the universe [as told in *The End of the Concept of Time*]; and its shocking, unbelievable consequences regarding the rest of the construction of the cosmos. Year after painful year, every time I wanted to quit, to give up, I remembered that awful day I had to quit the UDT; then I somehow dredged up my courage again, dragged myself out of the depths of despair I was feeling, and continued on and on; until finally, one day seven years later…I succeeded!!! [This mathematical analysis later became the thesis for which the Russian Ministry of Education in 2005 subsequently awarded me a PhD in Nuclear Physics].

Chapter 3
Boolean Algebra: The Science of Logic…

Reflecting back about all of it now, it is almost as if fate was training me slowly, step by step, giving me the emotional strength and the scientific knowledge that would be needed later on in life to discover the magic key, the Rosetta Stone needed for unraveling all of the great mysteries of the universe. Perhaps the one piece of knowledge that helped me the most was Boolean Algebra.

After I had to quit the UDT, I was sent to the Navy's Cryptographic school in Portsmouth Virginia. This secret school taught you how to repair the two most complex pieces of electronics equipment ever developed: the KWR-37 receiver; and the KW-7 transmitter. These two pieces of electronics were used by the United States Military and State Department Embassies all over the world to send and receive top secret messages. Because they are obsolete now, I am able to talk in generalities about them without fear of being arrested.

The key to understanding how these two wonderful pieces of electronics worked was Boolean Algebra. As we were told on day one in Crypto School, Boolean Algebra is the logic branch of mathematics that was used as a blueprint to create the electronic circuitry for these one of a kind machines. So, to be able to diagnose problems in the machines, one had to first become an expert in Boolean Algebra logic. Laying the foundation for becoming if you will, an Electronics "Detective" Technician with the skills of a Sherlock Holmes.

This is no exaggeration. To repair the equipment, one had to first diagnose the problem in these machines by having the attitude of a detective gathering evidence in a "crime scene." One did this by first examining the "print-outs" on the front panel, searching for any discrepancies; then finding and following the correct logic path in the manual's many schematics to the defective circuit that was creating the erroneous pattern.

This gathering of evidence so to speak became an important skill later on in life when I had to first spend years accumulating all of the major discoveries of all the important scientists for the past two hundred years; then take all of this knowledge and assemble it into a massive visual mosaic that created the new and revolutionary vision of the universe. A process that was all based upon the principles of Boolean logic. A procedure where the false and misleading information in current science – "erroneous information" – which was based purely upon supposition and assumption, and not fact; were first identified, then discarded, and not allowed to contribute to the end product of the mosaic [Such as Einstein's invention of a fourth dimension called "Space-time." Something which is *pure supposition*, based upon conjecture only, with absolutely no physical proof of its existence!].

It is an interesting anecdote to note that the Crypto school was so secret that note paper was officially numbered, assigned to each student, and locked in a safe at night. One student who mistakenly threw away a piece of this officially assigned note paper into a trash can, was arrested, given a court martial, and booted out of the service with a less than honorable discharge.

The manuals given to each student [upon which we marked the many different voltage lines with different colored pencils for quick identification] were kept under armed guard when not being used, and also locked in a safe in the school at night. When we graduated, these materials were then sent to our new duty stations in military vehicles and planes under armed guard. For a while, my manuals were kept in the safe in the Administration Building at the Charleston Naval Base in South Carolina where I was sent after graduating.

When I was called on to repair this secret crypto equipment on one of the many mine sweepers located there, my manuals and note papers were put into a special heavily reinforced and locked attaché case and given to a Marine Corps Officer carrying a loaded 45 caliber pistol, who handcuffed the attaché case to his left wrist, and escorted me to the ship [I had to walk on his left side, two paces in front of him when the pistol was on his right side]. He then sat in the doorway of the radio room watching me while I worked upon the equipment. I say all of this because it is almost laughable to note that later, we found out that the year before I attended this secret school, a Navy Chief Warrant officer named John A. Walker had sold both manuals to the Russian KGB! They actually had copies of the machines built and were trying to use them in their KGB offices in Moscow. In essence, all this secret security was a joke!

Luckily, these fantastic machines also used the 80 column IBM multi-punch cards. These IBM multi-punched cards used in the equipment had billions of combinations. They were also changed every 24 hours and the computers of that era were not capable of recording and later deciphering the massive streams of data the crypto machines sent out each day. Hence they could not play back the recording and then try some of the different billion punch card combinations in their machines to decipher the messages. It is also kind of humorous to know now that what the KGB bought for a lot of money was totally worthless without the cards! In essence, the conniving KGB got conned, by a con!

PART II
THE QUEST BEGINS

Chapter 4
The Near Death Experience in Readers Digest Magazine…

I was honorably discharged out of the Navy three months early in the winter of 1972 to be able to prepare for, then attend college in January 1973 at Orange Coast Junior College in Costa Mesa California. I had to start out in Junior College because my High School grades were deemed too low to be admitted to a university. But I am happy to report that unlike High School where my grades were so low I barely graduated, my experiences in the UDT trained me to study hard, to persevere and to win; giving me straight A's for this first semester. Then, after leaving for a while and after finally coming back to finish in 1975, I achieved a 4.0 Grade Point Average; was a member of Alpha Gamma Sigma, the California State Scholastic Society; and was graduated with honors in 1975 with an Associate of Arts Degree. My parents were shocked to learn that my scores on the tests I took were the highest ones in every course I took! But school at that time in my life was not meant to be.

A series of personal events caused me to leave school, move to Florida, and take a job repairing the massive telephone switching equipment then being used in the big hotels on Miami Beach. However, wanting to be outdoors, I soon quit, and accepted a job in swimming pool construction that paid more money. I applied myself; learned the skill of installing rebar in swimming pools; then became a sub-contractor and started working for myself.

And this is when my destiny began…and my purpose in life unfolded before me!

I must say here that I never intended to discover anything. If anybody had ever told me that one day I would just impulsively "up and go" to Florida to undertake a lifelong quest in search of the ultimate secrets of the universe, I would not have believed them. I certainly would not have believed them if they had told me I would eventually succeed and actually discover the knowledge sought from the beginning of recorded history by the world's greatest philosophers, sages, and scientists. And yet, it is exactly what happened.

It happened because of a mysterious incident that occurred one afternoon in 1974 in a small country store in Northern Florida. I use the word "mysterious" because over the years I have come to believe what happened on that day was meant to happen. That my arrival at the store on that day at just the precise moment was not just a coincidence. It was as if destiny had placed the right man, at the right place, at exactly the right moment in human history for whatever consequences await us all.

As many readers were probably too young to remember [or not yet born!], in the fall of 1974 the energy crisis sweeping across the country hit South Florida like a hurricane and the construction industry suffered greatly. Its timing could not have been worse.

I had just started working for myself in Fort Lauderdale as a sub-contractor in the swimming pool construction business. But when this mini-recession began, people had trouble getting loans to buy pools and I soon found there was very little work here or in any of the surrounding towns. Since I had spent most of my savings getting my business started and with no money coming in, I was going broke fast. So, after carefully considering my options, I decided to leave and look for work elsewhere. For some reason, I had always wanted to go back to a town in Northern Florida called Jacksonville where my ship had spent some time in 1971 while in the Navy. So, I packed my bags and went.

I drove the length of the state, and soon found a tiny apartment on Jacksonville Beach. I got a job on the night shift at a small fiberglass plant, which allowed me to pay my bills and to look for pool construction work during the daytime. However, after personally visiting all of the swimming pool companies in Jacksonville, I learned pool construction was just as slow here as it was in Fort Lauderdale. But that didn't stop me.

Undaunted, I next went to all of the surrounding towns looking for work. And when I didn't find any I tried other towns in other counties too. But no matter how hard I tried, I ended up with the same discouraging results. In one last desperate effort, I drove all the way to Tallahassee – the State Capitol – and spent the day contacting most of the swimming pool construction companies in the area. But it was not meant to be. The construction business was the same everywhere; there was no work available, period.

To say I was disappointed was an understatement. I felt like a fool for having come up here. And that night, as I sat in my hotel room, I wondered why I had ever come to Northern Florida.

The next day I found out.

As I was driving back to Jacksonville, for some unknown reason I decided to abandon the road I had driven the day before and take a more leisurely route through the countryside. Perhaps it was just an impulse on my part, or as others have since suggested – perhaps it was all part of destiny's plan. But for whatever reason, when I was about halfway to Jacksonville I turned off of the highway and began to drive along one of those old narrow country roads crisscrossing Northern Florida.

Even though this outdated collection of farm roads ran roughly parallel to the direction I wanted to go, after several hours of frustrating driving behind slower moving farm vehicles I began to wonder if my impulse was a wise one.

Thinking perhaps I was foolish to have taken these slower roads when I might have already gotten to Jacksonville, the sight of a small country store up ahead made me want to stop, get a soda, and take a short break. And that decision changed my life forever.

Just like a plot from a "Twilight Zone" episode, a meeting was about to take place in that little store that will one day affect the lives of everyone upon this planet. A statement which is even harder to believe because nothing about this old, dilapidated building seemed important, in fact, just the opposite.

The shingles were coming off of the roof, the long wooden porch running the length of the building was sloped inward, and the small gravel parking lot had deep ruts in it making parking difficult.

When I climbed the front steps they seemed to creak in protest, and when I opened the screen door, the rusty hinges screeched so loudly I tried to close it as gently as possible to spare the ears of those inside.

But the noise didn't matter. Except for a lone clerk who appeared to be intensely reading an article in a small magazine and didn't even look up when I walked in, the store was empty of people. Empty of people yet crammed full of everything else. Crowded into it was anything anybody ever needed. Packed onto the shelves, leaning up against the walls, and hanging from the ceiling were all sorts of farm tools, sacks of feed, work clothes, cans of food, soft drinks, and yellow straw hats. Everything was there, including that which I least expected to find – the keys to unlocking the secrets to the mysteries of the universe. (Quite a place now that I think about it.)

Although 47 years have elapsed since that incident occurred, perhaps its continued clarity is due to the fact I have never ceased to marvel at the incredible synchronicity of the events that happened on that day. Wondering over and over again what might have happened if I had not taken those particular roads. Or if I had left Tallahassee just five minutes sooner or later, or if there had been no slow moving traffic to delay my arrival at the store until precisely the exact time needed. Repeatedly I have gone through all of the "what ifs" that might have occurred which could have prevented me from walking up to the front of the store and arriving at the cash register, at the exact moment the young man finished reading the article.

This young man whom I never saw before and never saw again, looked up at me as if seeing me for the first time. He started exclaiming, "I just don't believe it. I just don't believe it."

Then totally ignoring the money in my hand, he started telling me about the bizarre story he had just read in the *Readers Digest Magazine,* which lay on the counter before us. For no apparent reason at all he started talking to me as if I was an old friend, showing me the article the magazine was opened up to.

Although I listened politely to him, my mind was on getting home. Even though I eventually told him I was going to have to be leaving, he continued to talk to me. He kept talking to me until I decided to leave. He kept talking to me all the way to the door. And as I started to walk out, much to my surprise, he grabbed my arm and insisted I take his copy of *Readers Digest* with me. In fact, he almost wouldn't let me out of the door until I accepted the magazine. (I left so quickly I forgot to pay him for the soda, and he forgot to ask me for the money.)

Perhaps he longed to talk to me because I was the only person in the place, or, as mentioned before, perhaps it was all part of destiny's plan. But whatever the reason, I was sufficiently impressed by this man's sincere amazement to read the article myself.

Here it should also be mentioned that even though *Readers Digest* is an excellent magazine, I hardly ever read any magazines at all. I never would have read this magazine and this particular article if it weren't for this man's insistence it was one of the most incredible stories he had ever read. And he was right.

It was one of the most incredible stories I had ever read too - which is an understatement, for it turned my life upside down. Nothing was ever the same again.

This first of its kind article told about what has since come to be known as a "Near Death Experience" of a man from New York City. It seems during the course of a heart attack suffered by this man, his conscious mind was not unconscious at all, as one would normally expect. Instead, he found his consciousness to be in another location away from his physical body; a strange location – where he found himself confronted by a grid-like barrier of light – unlike anything he had ever experienced before. But even stranger was the curious fact that when he was revived, he somehow felt disassociated from his physical body. He felt he was more like an observer, watching what his physical body was doing, rather than being the master of it.

Today, these near death experiences are reported so often they are no longer considered to be unusual or even unique, but, not in 1974. All those years ago, the general public had no knowledge of these incidents whatsoever. So, to read about one of them for the first time was a wonderful - yet puzzling experience. For even though the people whom I showed the article to were delighted by the spiritual implications of consciousness surviving death, at the same time they were baffled and incapable of explaining how such a strange phenomenon could occur: a paranormal phenomenon, which directly contradicted the theory of consciousness as proposed by western medical science. The theory that consciousness is a product of the neuron activity in the brain, and as such, is limited to the cavity of the skull.

Even though I was just as baffled as everyone else, what absolutely fascinated me was the fact that somehow there was a physical change in the location of the perspective through which this individual viewed the world. In other words, this person's conscious perspective of reality changed from the normal position within the body (just behind the eyes) to an abnormal position outside of it - something which is physically impossible according to our present knowledge of anatomy.

"However, the impossible cannot occur, only the possible can occur". Therefore, when the seemingly impossible does occur, it is the signal man's understanding of himself is either flawed or incomplete. That a new body of knowledge lies waiting to be discovered, which will revolutionize the way man views himself and his place in the universe.

I must admit, even though at that time in my life I was only vaguely aware of this truism, I was suddenly aware of a deeply felt need to understand, to know just how such an event could occur which directly contradicted everything I believed in. Even now it is hard to explain this feeling, but so great, so overwhelming was this need to understand, I was unable to think about anything else. For weeks, until I snapped out of it, I considered, rejected, reconsidered, then rejected many exotic electro-biological and psychological explanations.

For example, I thought it might be possible to create an upward extension of the brain's electromagnetic field away from the body. Or then again, perhaps somehow the unconscious mind of this individual was still active, still alert, and later fooled his conscious mind that wasn't functioning. That he was only imagining everything that happened. Again and again, I tried to come up with a physical explanation until exhausted – I was at last forced to accept the possibility such a phenomenon could also be explained by the existence of a soul within the human body.

That perhaps the change in the conscious perspective of this individual was caused by a premature release of the soul from the body when the body was in a condition that simulated death. That it was not the consciousness of the physical body that was observing what was occurring, but rather the consciousness of the soul. And that when the physical body was revived, and the soul was somehow drawn back inside, "its" memory of "its" experience was fed into the consciousness of the physical body which now perceived this experience to somehow have been its own.

This hypothesis was a major event in my life. I had been raised as an atheist, and had no religious beliefs whatsoever. I did not believe in a soul, I did not believe in God, nor did I even consider such beliefs to be worthy of any serious contemplation or speculation.

But now I could no longer ignore what this incident implied. Especially after being trained to use the reasoning process of the "scientific method" – which has as one of its cornerstone postulates the "uncompromisable", unyielding axiom that all possible solutions to a problem must be considered no matter how strange or implausible they may seem to be to the investigator. Looking back in retrospect, it is kind of odd, but it was due to the reasoning process of the scientific method I was now confronted with the real possibility the human soul actually existed.

Suddenly, I didn't know what to think anymore.

I was a 26-year-old man who grew up believing a man was nothing more than a biological accident. He was born, lived, died, and that's all there was. There was nothing more, and those who believed so were only fooling themselves. It was a tough time; but then it never is an easy time when an individual is rudely confronted with the unsettling probability that his beliefs are a sham, and that his most basic and fundamental thought processes themselves are built upon false premises.

Yet sometimes, what is originally perceived to be a disadvantage can actually be an advantage in disguise.

Sometimes adversity can transform itself into a powerful motivational force, creating an unrelenting mental drive that lifts and propels an individual above and beyond the confines of his former unfortunate predicament. A mental springboard whose impetus gives him a tireless self-motivation which drives him, and forces him to achieve much more than he ever could have accomplished when confronted with the mere presence of an ordinary set of circumstances alone. And so, it was with me.

My lack of knowledge of the subject of the soul, coupled with an intense need to understand everything about it, and hence fill some sort of emptiness within me, now became the motivating and driving force that compelled me to find out everything I possibly could about it. And I did. I did until a problem developed.

Much to my distress, a terrible emotional conflict began to grow within me. This conflict started out small and then began to grow larger and larger until it consumed my life. It set me upon the lifelong quest that culminated "ironically" in the greatest scientific discovery ever made. I say "ironically," because the conflict within me did not deal with science at all. Strangely enough, it centered itself upon one of the great philosophical mysteries of religion: the location of the Kingdom of God. If the Kingdom of God really exists – WHERE IS IT?

Chapter 5
Where Is the Kingdom of God Located?

The terrible conflict seething within me was a conflict many of you may have suffered yourselves. It was a conflict between the vision of the universe containing God, Souls, and the Kingdom of God, and the vision of the universe we see through our mightiest telescopes. A conflict between what can be called the "Religious Vision of the Universe" and the "Scientific Vision of the Universe". A conflict created by the following problem:

When the body dies, and the soul leaves the location of the physical body, it must travel to another location. Or rather, when the soul leaves the co-ordinates occupied by the physical body it once inhabited, it must travel to another set of physical co-ordinates. A set of co-ordinates designating a physical location which has to be *just as real* as the physical location occupied by the soul when the soul was in the physical body.

Since it is assumed this place is a Kingdom of Souls called various names in various religions, one characteristic common to all of these different beliefs is the fact that whatever you choose to call it, this kingdom has to be a real place existing in a real physical location. If not, this Kingdom of Souls is not a real place and doesn't exist. If it doesn't exist, there is a good possibility that neither the soul nor God exists either. And if there are no souls, and if there is no God, religion is a lie!

So where is the location of the Kingdom of God?

This problem began to trouble me, which seems odd, because the mystery of the invisibility of the soul didn't seem to bother me at all then (although it did later). At this time in my life, I believed in the tentative hypothesis the soul might exist as some form of "pure energy." Hence, the soul's invisibility would just be a natural consequence of its construction, and this rationalization seemed satisfactory.

But the mystery of the location of the Kingdom of Souls could not be rationalized away. So, much to my own chagrin, every time I found myself wanting to believe in the soul and in God, the mystery of the location of the Kingdom of Souls was always in the way.

Soon, the mystery of the location of the Kingdom of God began to occupy all of my conscious thought. All day long I would find myself thinking about it. The logic of the soul relocating to another position after the death of the physical body was irrefutable. No matter how hard I tried to rationalize it away, I couldn't do it. Nor could I ignore it.

To ignore it meant retreating into a fantasy world where I rejected the great discoveries of Astronomy, which was something I refused to do. All my life I had been fascinated by Astronomy and the latest discoveries of this subject were still fresh in my mind after having taken the latest college course in it before dropping out of school a year and a half before. So, I was left to suffer with this emotional conflict

But nothing lasts forever. To alleviate the anxiety I was suffering from, I realized I had to start taking some sort of action. So, on one particular evening, I deviated from the normal routine.

I carried a lounge chair out away from the buildings and the other people of the hotel. I hauled it all the way across the wide beach and positioned it down by the water's edge. And as the sun set in the west and the stars began to appear in the east, I decided to begin my quest for answers by defining the exact nature of the problem bothering me. A problem that begins and ends with credibility. For example:

The credibility of any religious belief that includes a kingdom of souls balances precariously upon a believable location for this Kingdom of God, or Heaven. Never before in the history of mankind has the knowledge of the location of the Kingdom of God been more important. Before the Seventeenth Century, this question was not a problem. Since early religions worshipped the Sun, Moon, and the planets, Heaven was always assumed to be somewhere up in the sky beyond the stars. Since nobody could see beyond the stars, nobody knew what lay beyond them. Hence, this explanation was satisfactory. Because nobody knew what was right, nobody could say what was wrong. Then came the invention of the telescope, and everything changed.

Four hundred years ago, Galileo turned his newly built telescope into the night sky and saw many wonderful sights, but he didn't see the Kingdom of God. This event probably surprised him as well as everyone else of his era - an unforeseen development that eventually presented the leaders of Christianity with a problem. Since everybody knew "Heaven" is above, and now that we have the means to see it, why can't we see it?

Although the priests of later eras were probably bewildered by this continued failure, the atheists were not. They possessed this powerful argument: *"If it is known to those who believe in Heaven that Heaven is above us somewhere up in the sky, and we have looked for it and cannot see it, it must not be there. Conclusion: the Kingdom of God is not real, it doesn't exist."*

In other words, since the *assumed* location of the Kingdom of God doesn't exist, it was concluded the Kingdom of God doesn't exist either. A convincing argument, even though it is based upon false logic. But false logic or not, since there is no other knowledge to contradict it, it has become an "acceptable solution" to an unwanted problem. Even though it is based upon a lack of proof, those who use it to justify their refusal to investigate further – accept it as a proof.

The reason why this argument is false is simple. To be able to say something exists is easy, all one needs to have is evidence of its existence. But before anyone can say something does *not* exist, they must know everything that does exist everywhere within the universe. The statement "something does not exist" is provable only by elimination. Elimination that can only be verified by a complete and thorough search of the entire universe. (Note, only after man explored the entire world was he finally able to state that dinosaurs were extinct, and even then, the prehistoric Coelacanth, a fish thought to be extinct for over 70 million years, was discovered in 1938 living in the ocean waters off the coast of Southern Africa.)

Consequently, before any man can state unequivocally, and with absolute certainty, the Kingdom of Heaven does not exist, he must first explore the entire cosmos before such a statement is valid. Without an exploration, such a statement is an erroneous assumption, an illogical opinion. An irrational opinion when one continues to believe it in spite of being made aware of the fact that it is illogical. But it gets worse.

There is another problem that contradicts the above logic. Amazingly enough, the discoveries of 20th Century astronomy now reveal a high probability the Kingdom of God *does not* reside within the physical universe observable through our telescopes.

This little known possibility must be revealed due to the fact that today, many well educated, and highly technically trained people of all different religions are trying to integrate their beliefs with the common beliefs of present day astronomy. Unfortunately, many people have reached the equally invalid conclusion that the physical matter of the universe might block the telescopic vision of the Kingdom of God. Or rather, we might not be able to see the Kingdom of God because something might be in its way, like a dust cloud, or another Galaxy, etc.

Although this argument sounds logical, it is probably wrong due to two reasons: the impermanence of the physical universe – as seen from man's observations; and the permanence of the Kingdom of God – as per the words of Jesus in the NEW TESTAMENT.

The impermanence of the physical universe is now one of the cornerstones of Astronomy. Although we see the same sky Aristotle saw, it is not the permanent place he believed it to be. Everything is in motion. Stars are now known to have life cycles. Some explode into Supernova, and some, like our Sun, will temporarily expand outward to a tremendous size. Someday, hopefully far in the future, our Sun will begin to expand until it reaches an incredible size, becoming a Red Giant millions of miles in diameter, vaporizing Mercury, and Venus, while boiling away the Earth's atmosphere and oceans. Turning the Earth into a blackened rock before the cycle is completed and the Sun collapses inward upon itself, becoming a tiny White Dwarf star.

Our Milky Way Galaxy is also in motion. Spinning like a gigantic pinwheel, this massive collection of stars moves through the universe with its companion Galaxies called the Local Group. Its fate is unknown. We know Galaxies collide, and we know Galaxies explode. Will ours? Some Galaxies also appear to have giant Black Holes at their centers, which are gobbling up their stars. Is this any place for the Kingdom of God?

Also, the universe appears to be the result of a cataclysmic explosion called the Big Bang. This belief in the "Big Bang" is backed up by physical evidence. The "residue" of the explosion is seen in the Cosmic Background Radiation, and the observation that all of the Galaxies possess a Red Shift, indicating they are all rushing away from each other.

Furthermore, if there is enough mass in the physical universe, it will end in what has commonly come to be known as the "Big Crunch", to disappear forever, or to begin the cycle all over again in response to the "Oscillating Universe Theory". A situation completely at odds with the words of Jesus, which states the Kingdom of God, will be forever. Which is the primary reason why the Kingdom of God probably does not exist within the physical universe; for how can something be forever amidst this maelstrom of change? It is unlikely. Therefore, it is also unlikely that the Kingdom of God resides within the confines of our physical universe.

This conclusion creates a dilemma. For even though the statement, "The Kingdom of God does not exist" is illogical, and even though it is unlikely that the Kingdom of God exists within the confines of the three dimensional physical universe – what is left? Is there any other place to look?

To one raised in a church environment and trained to believe in Jesus from childhood, such a question does not need to be answered. But to a former atheist raised in a scientific environment, it cannot go unanswered.

The mystery of the location of the Kingdom of God does not have to be solved by the individual who is "trained" to believe in God, because through the generations old process of memorization and recitation, he is literally programmed to believe what others tell him to believe. He is taught (by those who teach him), that those who teach him are teaching him "God's Will", and for him to then question them about where God exists, or where the Kingdom of God exists is to really question his own faith. For him to even ask such a question of them reveals there is something wrong with his beliefs. (Which is a very clever answer for those who do not have a clue themselves about how to answer such a question, and don't want to suffer the embarrassment and loss of credibility - and the challenge to their authority by being unable to answer it.)

Nor does the mystery of the location of the Kingdom of God have to be solved by scientists, because science does not acknowledge the existence of God. Only to a former atheist trained in the principles of the scientific method does it become a problem, a problem that does not go away. A problem that stays with you and eats out your guts until you have to do something about it.

When I first began to work upon this problem, I did not know if it was solvable. However, the more I thought about this ultimate puzzle of puzzles, [this conundrum], the more I began to realize the Kingdom of God's invisibility might be the solution instead of the problem, because if it cannot be seen, there might be a reason why it cannot be seen. We might be looking for it in the wrong place. For if it really exists, yet doesn't exist within the physical universe, it must be located someplace else; it must exist in another location where it can be just as real as we are, yet remain totally invisible to our eyesight.

Is there such a place? I had to find it. The search for this answer now meant much more to me than just finding the solution to a question and satisfying some intellectual curiosity. It was the search for the purpose and meaning of life itself. For if the soul does indeed exist, and the Kingdom of God is real, then we aren't just a bunch of biological microbes living upon a speck of sand in the middle of a cosmic maelstrom, as some have insisted. Instead, it means there is a purpose for being alive.

So, what is this purpose I asked myself, and what is its importance?

The answer I came up with goes like this: if life has meaning, then the purpose in life is to find that meaning, but if life has no meaning, the search for a meaning gives one the purpose in life they would otherwise not have. So, either way, the search for the meaning of life – gives meaning to life.

And as I got up from my lounge chair and stood under the brightening sky proceeding the dawn, I vowed to give meaning to my life. From this day forward I would have a purpose in life, not just a vain purpose in pursuit of an impossible goal, but a real purpose. To actually succeed, to do what no man had ever done before, to solve the mystery of the invisibility of the soul, and find the actual location of the Kingdom of God.

Although the above statement might sound like some sort of overconfident "bravadoism," one thing I did not lack was confidence. I had supreme confidence in my abilities. A confidence, born of success.

As mentioned earlier, during my military service in the Navy, I was trained to be an Electronics Cryptographic Technician. I was trained to seek solutions to very complicated problems using Boolean Algebra. A job leaving me skilled in finding problems in some of the most sophisticated electronics equipment ever developed. Hence, I believed in myself, and I knew if I could find the solutions to troubles in very complex electrical circuitry, I could also find the solution to the mysterious location of the Kingdom of God.

But I was also a realist. Even though I was confident and full of spit and vinegar, I knew my ability in solving electronics related problems was a direct result of the knowledge I possessed about electronics. While the problem I was now trying to solve required a completely different type of knowledge. What type of knowledge was necessary to solve this seemingly impossible problem of problems?

There was only one solution I could think of: school. Go back to College. There was no other choice. So, in December of 1974, I packed my bags and headed west to California.

Chapter 6
The Search for the Kingdom of God Begins…

In order to discover the mysterious location of the Kingdom of God, I realized I needed to do two things: I had to make an overall study of both science and religious philosophy, and I knew I must think differently from everyone else.

Although I wanted to learn what others knew, I did not want to become another victim of their mistakes. Therefore, I didn't want to think like they had thought. The reason why I didn't want to think like others was due to the observation they had not solved this problem for themselves. Hence, there had to be an error in their way of reasoning. An error keeping them from breaking the impasse and making the revolutionary discovery necessary to reveal the truth.

The key to making revolutionary discoveries lies in thinking differently from everybody else. It is not hard to understand that when everyone thinks alike, everyone reasons alike, and everyone comes to the same conclusions. When this erroneous reasoning process is passed on from one generation to the next for countless eons, no progress is ever made. It continues unless the cycle is broken.

I knew the only way to break this constantly repeating cycle, and reach new conclusions nobody has ever considered before, was to throw away the ideas of past generations and begin thinking the thoughts nobody has ever thought of before. Consequently, to think the thoughts no one has ever thought before, I knew I must first examine the facts upon which the current "thought system" of beliefs are based and ignore any opinions others might possess. After I had accomplished this goal, I was then going to try to discover if there is an alternative explanation that explains the facts just as well as the explanation currently accepted by everyone else.

A good example of the wrong system of thought people can be drawn to and entrapped in for generations, is the vision of the universe where it was once believed the Earth was located at the center of the cosmos. Even though we know today this idea was wrong, it nevertheless survived for thousands of years.

The reason why it survived for thousands of years was due to the fact that each generation continued to reason like the previous one. Each succeeding generation would observe the same motions of the Sun, Moon, Planets, and Stars seen by all previous generations. After seeing what everybody else previously saw, the reasoning processes they inherited caused them to agree with the explanations provided them by their forbears and ruling elite – that everything appeared to be circling the Earth. This idea seemed to be so easily "re-confirmable" for each new generation that came along, that everybody kept on thinking exactly the same way as everyone else before them. Except for a near-contemporary of Aristotle in Athens named Heracleides Ponticus, who believed the Earth was rotating upon its axis, very few people in history up to and including the era of Copernicus, dared to challenge the explanations provided them and think differently from everybody else.

This type of stereotyped thinking is what I wanted to avoid. So, when I finally arrived in California, my mind was made up not only on what to study, but how to study - to study the facts and avoid the conclusions made by everybody else.

Since I had already spent that one semester at Orange Coast Junior College in Costa Mesa, California after mustering out of the Navy two years before, I decided this was the place to return. So, in January of 1975 I re-enrolled in the spring semester at Orange Coast Junior College in the town where I had grown up. Since I was a Viet-Nam era Veteran, the government paid for my education, and I soon became a professional student; I studied and got paid for it.

Unlike other students, because there was no curriculum given for what I wanted to study, and since I hadn't told anyone what I was doing, I decided to start out by taking courses in Religious Philosophy and Science. Although, I already knew no school in the world possessed the knowledge of the location of the Kingdom of God, I was hoping to find some sort of clue which might help me in my quest. Since I didn't know where to look, or even what to look for, these courses seemed like as good a starting point as any.

Remembering how a previous course in Astronomy gave me the insight into the impermanence of the universe, and allowed me to realize the observable universe was no place for the Kingdom of God, I logically reasoned other science courses might give me other insights as well. Therefore, while at Orange Coast College, and then later, after graduating and transferring to California State University at Long Beach, I took courses in Geology, Astronomy, Physics, Chemistry, Archaeology, Anthropology, and Calculus.

Unlike the unmotivated years of my youth, because I now had a purpose to study and a reason to learn what I was studying, much to my surprise, I did very well. I got top marks, and as mentioned previously, I graduated from Orange Coast Junior College with a 4.0 grade point average, which wasn't bad for a kid who barely graduated High School. I also found I liked the academic environment at Cal State Long Beach. I excelled in both Calculus and Physics, finding I had a talent for these subjects. But amidst all this success, I was deeply troubled.

Since I was still caught up in the struggle between the scientific vision of the universe and the religious vision of the universe, my mind continually oscillated back and forth between these two seemingly contrary sets of ideas. So, whenever I could, I supplemented my college education by visiting many of the different religious organizations located throughout California. Since I looked at different religions as one might consider different solutions to a mathematical equation, I held no emotional attachments to any religious belief. Therefore, I studied them all.

I studied Hinduism, Buddhism, Taoism, Confucianism, Zoroastroism, Judaism, Christianity, and Islam. I studied the *Upanishads*, the *Tao Da Chang*, the *Batavia Gita*, the *Old Testament*, the *New Testament*, and the *Koran*. I studied the lives of Buddha, Lao Tse, Confucius, Zoroaster, Moses, Jesus, and Mohammed.

I practiced meditation and visited Monasteries and Ashrams. I conferred with all sorts of Priests, Monks, Rabbis, Swamis, and Theologians from practically every sort of religious discipline there was, but I was not at ease. In spite of everything I was doing both in and out of school, I felt I was making no progress towards accomplishing my goal. All I was doing was passing courses, making good grades, and yet somehow feeling empty because of it. The problem of the location of the Kingdom of God still eluded me. I felt deeply frustrated knowing I was no closer to the truth than when I started.

Since I was not at ease, I felt I needed a change. I was tired of living in the city. For some reason I decided I would like to go to Northern California, to the Great Redwood Forest. I had always enjoyed these majestic woods located north of San Francisco and just inland from the Pacific Ocean. As a child, my parents used to take me there upon their yearly vacations. I remembered a magnificent grove of Redwoods called the "Avenue of the Giants" and treasured the peace I felt while standing in the presence of these wise old trees.

So, recalling what happened when I went to Northern Florida and ended up at the little store, I felt some impulse was moving me again. Therefore, I went to the university library, read the brochures on all of the schools in the California State University system and discovered California State University at Humboldt. A school located in Arcata, a town right in the middle of the beautiful coastal rainforest. Since this seemed to be the school I was looking for, I went back to my lodgings, packed up my bags and headed to Northern California. And just like before, I was right on schedule.

Chapter 7
Higher Dimensional Space…

Although California State University at Humboldt is one of the smaller schools in the California State University system, it is one of the most beautiful. Located far up on the Northern California coast in the little town of Arcata, it sits right at the edge of a magnificent forest of giant trees. Called the "Backpackers School" because many students go there to study forestry, I myself temporarily considered studying forestry with the goal of becoming a ranger naturalist, something I always wanted to be as a child.

But it was not meant to be, because it was not the reason why I had come to this school. Fate had other plans for me. I came to this school to take a brand new course just being offered for the first time. A revolutionary physics course containing within its curriculum - the solution to the mysterious location of the Kingdom of God.

Of course, I knew none of this. Before I arrived at the school, I had gone through the catalogue and picked out the courses I intended taking. However, I was still curious about the rest of the courses not in the catalogue, yet being offered during this particular semester. So, as I idly looked through the list handed to me at the administration building, my interest in physics made me note the intriguing name of one particular course I didn't recall seeing in the book: "*Space-time Physics*". Asking about it, I discovered it was a newly developed course being offered for the first time. A course where Einstein's Theory of Relativity would be discussed along with other Physics subjects of current and curious interest. Although this course was not associated with any of the subjects I intended taking, I thought, "What the heck, it's only a two unit course and it won't take up too much of my time, so I might as well try it". So, I enrolled and unknowingly changed my life.

This new course was exactly what I had come to college for. It covered many of science's most intriguing ideas, including speculation about higher dimensional space. The theoretical possibility of the existence of fourth, fifth, and sixth etc. dimensions of space; unseen volumes of space existing at ninety degree angles to our own three dimensional universe: other universes.

I was never more alert. Unfortunately, nothing much was known about higher dimensional space at that time, so nothing much was said. Also, the existence of higher dimensional space was not considered to be a serious subject. Since the construction of a fourth dimension made of space alone conflicted with Einstein's fourth dimension made of both space and time, the existence of all higher dimensions made of space alone was considered to be highly unlikely. Hence this subject matter was only briefly mentioned, then passed by - but not by me.

This was the first indication I had that there might actually be a real location for the Kingdom of God outside of the physical universe. So naturally, I decided I needed to learn more about higher dimensional space: and I did. For the next few weeks I totally forgot about school. I researched all of the literature I could find in the school library upon this subject, which was a disappointment, because there wasn't very much at all available.

However, what I learned was enough.

Roughly speaking, higher dimensional space is another volume of space located outside of our three dimensional universe. This volume of space has a unique perspective: an observer in three dimensional space cannot see anything existing in higher dimensional space, however, from higher dimensional space, an observer *can see* everything existing in three dimensional space.

I was fascinated. At last, here was the realistic perspective fulfilling one of man's most universal and widely accepted beliefs about the omnipresent nature and ability of God - from the perspective of higher dimensional space, God can see everything and know everything happening within our three dimensional universe, yet not be seen!

Higher dimensional space also fits the requirements for the location of the Kingdom of God. From higher dimensional space the Kingdom of God can actually exist in a real physical location. (A real physical location, yet invisible to everyone in three dimensional space.)

Higher dimensional space is also the perfect location for the soul. If the soul is a higher dimensional creature, even though it is a part of us, it too will be invisible. Although it is hard to believe [but true], the soul can be but a millimeter away from the geometric center of the physical body, yet never be seen!

I was ecstatic. Higher dimensional space seemed to fit all of the requirements necessary for the existence of God, the Kingdom of God, and the soul. Or rather, to say I was ecstatic was an understatement. I could not express the emotional euphoria I was feeling. To dream the impossible dream of accomplishing the impossible goal is one thing, but to actually come across a possible solution created an emotional rush that flowed like a river of life through my body. Every cell was supercharged with energy. I was reinvigorated, refreshed after the long journey. I felt whole again, strong again.

It was a great time to be alive. Higher dimensional space had to be the location I was looking for, for if it exists, it suddenly provided an explanation for many of the ideas about religious philosophy which were only assumptions the day before. However, the key words are, "If it exists". For in Einstein's vision of the universe there are only three dimensions of space, and one dimension of time. There is no fourth dimension made of just space alone.

Or is there?

How many times in the past have men rejected an idea because it didn't fit in with the popular beliefs of their era? Looking back at history, how many men two thousand years ago would have believed that the world was not flat, or that the world did not have to be supported by something else [like a giant turtle (India, China, American Indians); or Atlas (Greeks)]? Who in those days would have believed the world was really a sphere hurtling through a dark void of unimaginable size? A void so vast that if it could be compared to the size of an ocean, the Earth wouldn't even be equivalent to the size of a grain of sand. No, I doubt that anybody then would have believed such a tale. And yet it is true.

So, armed with the knowledge the true reality is often more bizarre than the reality we envision, I began to spend all of my time examining the unusual qualities and properties of higher dimensional space. In fact, I spent so much time thinking about higher dimensional space, I had to change all of my courses from "credit" to "noncredit" to keep from destroying my grade point average and flunking out of school.

I must admit I was uneasy knowing I was deliberately neglecting the future a college education could provide me with by spending all of my time on this seemingly frivolous pursuit, but all of this extracurricular activity eventually paid off. It paid in a way I would never, ever have anticipated – with a spectacular insight into the words of Jesus in the Gospel of Luke. An insight unlike anything anybody has ever glimpsed before - a revolutionary insight bridging the gap between science and religion.

It all happened one morning while strolling through the beautiful forest which grows right up to the edge of California State University at Humboldt. For some reason I had decided to take a morning walk. I had walked across the campus, entered the forest, and was following a narrow winding trail weaving in and out beneath the massive trunks of these majestic trees. Although surrounded by the beauty of nature, my mind was elsewhere. As usual I was thinking about higher dimensional space, and on this particular morning I was trying to envision the entrance into and out of higher dimensional space.

The entrance into higher dimensional space is a most unique motion - unlike any other motion in our three dimensional universe. It is difficult to imagine because there is nothing to compare it to. To enter into higher dimensional space, one has to travel at right angles to all three dimensions of space simultaneously. If that doesn't seem to make much sense, the only way to explain this direction is the use of the word "within": within space itself. Or, when matter is used as the reference point, "within" matter itself - in a direction towards the geometric center of matter.

If you seem a little confused, don't feel bad. Since we think and reason in terms of three dimensional space, higher dimensional space is almost impossible to visualize. This problem occurs because of the way higher dimensional space is constructed. Higher dimensional space is a volume of space at right angles to the volume of three dimensional space we inhabit.

A way to envision the relationship between the different dimensions of space is in the following drawings:

Figure 7.1 One dimensional space consists of one line.

Figure 7.2 Two dimensional space is a plane that is at right angles to the line.

Figure 7.3 Three dimensional space is at right angles to the plane.

Figure 7.4 Fourth dimensional space is at right angles to three dimensional space at once! Since Fourth dimensional space is impossible to draw, the following is only a representation, a schematic drawing.

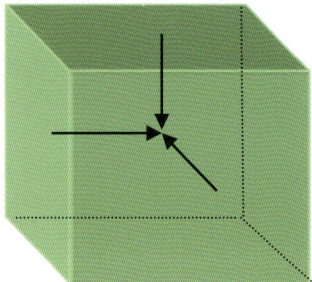

Another way to understand this construction is to look up at a corner of the room you are sitting in. Observe the point where the ceiling and the two walls meet. The ceiling and the two walls represent the three dimensions of space we live in: length, width, and height. Now observe how the ceiling is at right angles to both walls, while both walls are at right angles to the ceiling and the opposite wall. This relationship allows us to observe the fact that each dimension is at right angles to the other two dimensions *simultaneously*.

Although this three dimensional relationship is easy to see, the relationship between these three lower dimensions and higher dimensional space is not. It is easy to say the fourth dimension is at right angles to the first three dimensions, and fifth dimensional space is at right angles to the first four dimensions and so on, but it is impossible to visualize this relationship.

What we can visualize is movement into or out of the fourth dimension. To do this, again look up at the same corner of the room and observe the tiny point where the ceiling and the two walls meet. To reach the fourth dimension we must travel within this point, towards its geometric center, and then outward, into the fourth dimension.

Although this doesn't seem to make too much sense, it does when we realize every invisible point of space within the room you are sitting in is exactly like the point you are looking at. Hence, to reach higher dimensional space, one has to travel within space itself. Or, since every object is made up of an infinite number of such geometric points, a way to describe the location of higher dimensional space to someone who does not comprehend what you are talking about is to use the word "WITHIN".

As a demonstration, you can hold up a small pebble and say to a group of children that the location of higher dimensional space is "within" this very pebble. However, even though this is a dramatic demonstration, it is a poor illustration. Although what you said was true, the children might misunderstand what you are saying and think higher dimensional space is located within this particular rock. So, an even better way to describe the location of higher dimensional space is to use another illustration. For example, you could say higher dimensional space was not only within them, it is also within each and every one of us. By using the matter out of which our bodies are constructed as a reference point, we must travel within - within our own selves: towards the geometric center of our own physical body.

Wait a minute!…What the heck!…The geometric center…Yes! That's it!…The geometric center!

Suddenly I halted in the middle of the trail. Desperately, I found myself trying to recall something Jesus said about the Kingdom of God. Could it be possible? Had my thoughts about higher dimensional space revealed a startling revelation? Could there be a connection between the science of physics and the words of Jesus in the NEW TESTAMENT?

The idea seemed too incredible to be true. I wanted to reject it outright and immediately as nothing more than foolish speculation. Yet I seemed to vaguely recall something Jesus said - a quote in the Gospel of LUKE about the location of the Kingdom of God. For no apparent reason at all, this quote just popped into my mind: just as suddenly, I was running down the trail.

I was never a good runner, but on that day I was never so fast. I ran like the wind - back through the woods, across the campus, through the streets of Arcata, on the sidewalks and off, dodging in and out of traffic like a crazed wild man until I finally made it the mile and a half back to my lodgings at the old milk factory. With my chest heaving up and down, gasping for breath and the sweat pouring off of me, I finally got the fumbling key into the front door lock, and without bothering to close the door, dashed down the long hallway to my room.

Without stopping I upended and dumped the contents of boxes of books and papers all over the floor in a frantic search for my Bible - which, in my excitement, I forgot was still in the drawer of the nightstand beside the bed. When I remembered, I tore open the drawer, snatched up the Bible and practically ripped out the pages while turning them as rapidly as I could, searching for the quote I was trying to recall. The only quote made by Jesus in the NEW TESTAMENT describing the actual location of the Kingdom of God.

Then suddenly, there it was before me. I stopped turning the pages as my gaze fell upon the red letters of Jesus' quote in Luke 17:20-21. Stunned, I slowly sat down in my little chair. Even though I had read this quote before, I had never understood it. To me, and I know to many others, this quote always seemed to be just another parable referring to the heart or the mind. However, I now knew this wasn't true. Suddenly I knew it meant something more. [Something much more!]

Just how much more I had no idea. Little did I know then this seemingly insignificant little quote is the philosophical key to understanding how the entire physical universe is constructed. Possessing astounding implications, it is the key to unlocking the great scientific secrets of the universe that have mystified the greatest scientists who have ever lived. Secrets such as the particle and wave theory of matter and energy, the unification of the forces of nature, and many, many more.

Chapter 8
The Greatest Story That's Never Been Told…

When I first read the words of Jesus in the NEW TESTAMENT, there were a number of statements and parables I did not understand. But there were also a few which seemed to be so obvious, no other explanation appeared necessary. One of these was Luke 17:20-21. Since Jesus made so many references to the heart and the mind, this statement just seemed to be another parable referring to these human attributes. In fact, I do not ever recall even questioning this explanation. Nor do I know of anyone else who ever questioned it either.

But now, as I sat in my makeshift room at the old milk factory in Arcata, carefully reading and re-reading LUKE 17:20-21, I realized this quote did not refer to the heart or mind at all. I knew then it meant something more, something much more, and oddly enough, I found myself scared.

I was scared knowing I, and I alone was privy to the frightening truth that 2000 years ago, Jesus knew more about the construction of the universe than all of the greatest scientists who have ever lived. I use the words "frightening truth", because at that moment in my life I suddenly knew Jesus possessed knowledge far exceeding anything any man has ever known. How he got it, I did not know. But what I did know was somehow, he was tuned into a source of knowledge that far surpassed the scientific knowledge of any era in human history including ours.

I say, "Tuned into", because he did not get his knowledge from anyone else, "nor" did he get it through experimentation. So, what was his source? Was it God? If it was, then without a doubt, Jesus was the Messiah: the messenger of God whose coming to Earth was foretold of in the *OLD TESTAMENT*.

I was overwhelmed. I felt privileged and honored, yet strangely enough, unworthy. At that most precious moment of all moments, all I could think about was all of the bad things I had done in my life. There were so many others in the world who were much more worthy than I. I was disappointed with my failings; I lay back upon my bed weeping and suffering great anguish.

Sometime during that afternoon, I must have fallen asleep. Later, I awoke in the darkness, realizing I was hungry, yet not caring. Instead, I turned on the light, and opened up the Bible just to look again at what now seemed to be the most important of all the quotes made by Jesus in the NEW TESTAMENT. Like a prospector who has just discovered a great diamond hidden in the desert, I had to keep looking at it again and again to make sure it was real, and I wanted to be close to it to touch it; to hold it; to cherish it.

And when he was demanded of the Pharisees, when the kingdom of God should come, he answered and said, **"THE KINGDOM OF GOD COMETH NOT WITH OBSERVATION: NEITHER SHALL THEY SAY, LO HERE. OR LO THERE. FOR BEHOLD. THE KINGDOM OF GOD IS WITHIN YOU."** *LUKE 17:20-21*

As you read this quote, try to imagine you are seeing it for the first time. I know for many of you who are Christians it is difficult to ignore the training of a lifetime, but don't cheat yourself. Like everyone else who is not a Christian you also have a right to know the truth.

Although most Christian peoples are trained from early childhood to believe this statement is a parable or metaphor referring to the heart or mind, is it just a coincidence it is also an exact description of the location of higher dimensional space? By using physical matter (the physical body) as a reference point, the direction of higher dimensional space (if it exists), is "within", towards the very center of matter itself.

It is also important to understand that if Jesus used another reference point it might cause a conflict. For example, if the physical matter of the Earth, Sun, or Moon was used as a reference, it might create a misunderstanding and lead the listener to believe the Kingdom of God was located within the center of one of these celestial bodies. Or, if he had held up a stone and said the Kingdom of God was within, people might believe the kingdom was within that very piece of rock.

This word "within" is also a very important word. Even though other words are adequate such as "inward", or "inside", the word "within" does an excellent job of representing both the direction and the location of higher dimensional space. However, as before, the three most important words are "if it exists". For in the present view of the universe with only three dimensions of space and one dimension of time, there is just no room for a higher dimension of space made of space alone. Furthermore, ideas and speculations about the existence of higher dimensional space are legacies of the modern era of mathematics. During the time of Jesus, higher dimensional space was unknown. But maybe not to Jesus!

When Luke 17:20-21 is analyzed from the point of view of being a direct answer to a direct question, its implications are too intriguing to ignore. The previous misunderstanding of this quote being a parable referring to the heart or mind is easily understood. Unless one knows something about the geometry of higher dimensional space, this statement appears to be some sort of philosophical reference being made to thoughts or emotions. But this is not so.

To correctly analyze this statement, we must first realize this quote is a direct answer being made to a direct question. Consequently, it is not a philosophical reference being made to the heart or the mind, or as to how one thinks or feels. Nor is this statement a simile or a metaphor.

To correctly understand this statement, it must be remembered the Pharisee's believed the Kingdom of God would be an actual kingdom located upon the Earth. A belief contrary to the teachings of Jesus as demonstrated in John 18:36 *"MY KINGDOM IS NOT OF THIS EARTH"*. Consequently, in response to the Pharisee's question, the first words Jesus responds with, *"THE KINGDOM OF GOD COMETH NOT WITH OBSERVATION,"* are referring to this erroneous belief held by the Pharisees, and Jesus' purpose in answering this question is to correct this error.

Jesus then goes on to say, *"NEITHER SHALL THEY SAY, LO HERE, OR LO THERE."* In other words, the kingdom of God *cannot be seen with, or discovered with one's physical eyesight.* (Note: the word *"LO"* denotes exclamations of discovery.) Finally, he ends this statement by boldly stating, *"FOR BEHOLD, THE KINGDOM OF GOD IS WITHIN YOU."* Although these two statements are short and simple, they are most profound. For if the Kingdom of God cannot be seen with one's physical eyesight, it does not exist within our three dimensional universe. Which means the only place it could exist would be in another dimension of the cosmos.

Because all dimensions are at right angles to each other, the only way to enter the fourth dimension when using matter as a reference point, is to travel at right angles to all three dimensions simultaneously, or "within" - towards the very center of matter itself. The exact direction Jesus is alluding to when he states, ***"FOR BEHOLD, THE KINGDOM OF GOD IS WITHIN YOU."*** He adds special emphasis to this statement by adding the word, ***"BEHOLD"***: which means - "look", "see". A word he seldom used except to emphasize an extraordinary statement - a revelation.

Here we must pause and ask ourselves if we are expecting too much of Jesus. The concepts of higher dimensional space belong to Twentieth Century cosmology, not to ancient religious philosophy. Or do they?

If Jesus was answering an erroneous question with an accurate view of reality, he already knew about higher dimensional space and didn't have to wait for modern man to invent it.

If this analysis is indeed correct, if it is in fact a correct explanation of Luke 17:20-21, its implications leave us stunned. If the Kingdom of God does indeed exist in higher dimensional space, it means higher dimensional space exists, and if it exists, the present vision of the universe with its three dimensions of space and one dimension of time is in error. Hence, it means that if by using the words of Jesus in the New Testament we can find that error, correct it, and discover the true vision, the course of human history will be altered.

Although Christian peoples already believe Jesus as the Messiah "knew things no man has ever known", for the first time ever, the rest of the peoples of the world will know it too. They will have scientific proof Jesus' knowledge of the construction of the universe exceeded and surpassed the knowledge discovered by all of the greatest scientists who have ever lived.

It must also be emphasized, the knowledge of the existence of higher dimensional space is not just any knowledge. It is very difficult – extremely difficult knowledge to obtain. This knowledge stands upon the apex of a pyramid of other scientific discoveries. It is knowledge obtained only by the most precise and exact experimentation.

Even more important, it will soon become apparent higher dimensional space is not just some mundane, insignificant place. Quite the contrary! Higher dimensional space is the most important place in the cosmos. Without higher dimensional space it will be shown that nothing can exist within our three dimensional universe. Without higher dimensional space, we cannot exist.

In knowing all of this, can there then be any doubt Jesus was the Messiah? Will the attention of the people of the world shift towards this most wonderful man of all men? Can a renaissance in Christianity be created?

The stakes are high. Can it be done?

Can the course of human history be altered by one quote from Jesus?

Yes it can, if the error in the scientific vision of the universe can be found.

Chapter 9
Somewhere There Is A Problem With the 20th Century Vision of the Universe!

After discovering the correlation between the words of Jesus in the New Testament and the theory of higher dimensional space, school was never the same again.

I knew if higher dimensional space existed, man's vision of the construction of the universe was wrong. Furthermore, since the error dealt with the construction of space, there also appeared to be an error in the Theory of Relativity - for it was Albert Einstein who developed the current view of space; a deduction which meant trouble.

In the world of science, Albert Einstein has achieved superstar status. An icon of his era, he has a following comparable to that of a cult leader. Even amongst those who have never studied any science whatsoever, the mere mention of the name Einstein conjures up thoughts of great genius and enormous intellect. If science was a religion [and to some it is], Einstein would be one of its greatest saints, and his Theory of Relativity would be one of the favorite chapters in its bible.

Unlike different religions with different sets of beliefs, the same science is believed throughout the world. Almost all scientists, in all cultures and countries accept the Theory of Relativity. Since scientists who believe in Einstein are responsible for training new scientists, nearly all of the scientists in the world end up thinking alike. Consequently, just about every physicist upon this planet sees a mistake in the Theory of Relativity to be a physical impossibility. In every college and university classroom, there is nothing but adulation and praise for Einstein. So unfortunately, within this climate of constant acclaim and approbation for Einstein, for a professor to speculate out loud to his students that Relativity might be wrong, is an act equivalent to the commission of "scientific heresy" in the academic world.

And yet, if higher dimensional space exists, man's current vision of space *is* wrong. Which means Einstein is wrong, the Theory of Relativity is wrong, and Einstein's vision of the universe is wrong. Unfortunately, Einstein's vision of the universe cannot be repaired with a "Band-Aid." The extra dimensions of space cannot just be added to the dimensions already there because there is no sound scientific reason for doing so. Such a change would only be a philosophical one. It would be meaningless because it would not be backed up by scientific proof.

So, I had a problem, I needed proof.

Without proof, I was painfully aware that the existence of higher dimensional space was nothing but supposition and conjecture. However, what gave me hope was the knowledge that if the location of the Kingdom of God really does exist, it will be a provable scientific fact. It will be provable using the techniques of the scientific method: the same method that was used to develop the current scientific vision of the universe. A vision so formidably ingrained into the minds of the present generation that it could only be changed by the same method that created it.

But where do you begin? How do you challenge a theory all of the world's teachers believe in and admire? How do you find a fault in the greatest theory in all of science? Not just any theory, but a theory developed by one of the greatest thinkers who ever lived. A thinker, who may just have been the very best of the very best ever.

Wow! Knowing what I was up against gave me no comfort. It wasn't going to be easy. I also knew I wasn't going to get any help. I was painfully convinced of that after casually mentioning to a professor of physics I believed something appeared to be wrong with Einstein's construction of space.

If you try it yourself, you will probably experience something like this: first, you will be looked at in disbelief; then suspiciously, like you are crazy or on drugs. Then, after a slight pause, you will be rudely told, "What can an undergraduate student [like you] think they can find wrong with the most difficult and complex theory in all of science - a theory which sometimes takes years to fully comprehend. Above all, how could someone like you, with hardly any training in the science of physics, prove what thousands of the most learned physics professors and researchers in the world can't prove"?

[And most amazing of all, this logic sounds perfectly reasonable.]

I knew it was difficult to oppose such a powerful argument as that one and I realized my prospects of finding something wrong with the Theory of Relativity, and discovering the true vision of the universe didn't look too good. I was in a David versus Goliath battle. But David found a way to win, and while thinking about a way to win, I realized I had one chance too. A slim one, but a real one. I knew from studying history I needed to find some sort of problem with the current scientific vision of the universe. If I could find such a problem, it would indicate the present scientific vision of the universe was inadequate. Being inadequate; it would then become apparent to everyone that it needed to be modified or scrapped altogether.

But could I find such a problem? Where should I begin?

Reflecting back upon the chemistry and physics courses I had already taken, I couldn't seem to come up with anything. But this initial failure didn't discourage me. I knew some sort of problem had to be there – there was no doubt about it. Again, if higher dimensional space did indeed exist, then the present scientific vision of the universe had to be wrong. It was just as plain and simple as that. But what was wrong? Even more important, how could I, a lowly student, discover what has been missed by all of the hundreds of thousands of scientists, professors, and graduate students throughout the academic system of the entire world?

While analyzing this problem, I decided that since I did not even know what to look for, my first task was at least to know where to look. So, I decided to reduce everything in the universe down into its most basic and fundamental parts. I called this "the scientific vision of the universe" – a term used throughout the remainder of this book – and found it consisted of only five things: matter, space, time, energy, and the forces of nature.

Next, I reviewed each part individually looking for any sort of problem… However, I could not find a problem!

[Note: although this was an important review, it is not necessary for our story. The review can be found in Chapter 6 of Book 1, *The End of the Concept of Time*.]

Chapter 10
Hard Times…

Identifying a problem is one thing, solving it is another.

Although I wanted to believe higher dimensional space existed, I knew I could not believe in it until I could prove it scientifically. But to prove it scientifically I had to find the mistake in the present scientific vision of the universe. This mistake had to be a major one, a mistake of enormous proportions. But at the moment, this mistake eluded me, and my failure to find it frustrated me to no end.

This frustration created a great deal of stress. The search for the mistake in the scientific vision of the universe was now one of the great quests in my life, but I was also a realist. I knew it might take years of schooling before I was able to discover this mistake, and as mentioned before, I had no illusions about getting any help from Physicists.

It was a tough time. I didn't know what to do. I only knew I wanted to learn the truth, and I now knew I could not learn the truth in school. This attitude made it hard to sit in class anymore.

Being absolutely certain a great error of enormous proportions existed somewhere in the scientific vision of the universe; it was hard to take seriously any of the science courses I was now enrolled in. Nor could I fool myself. If higher dimensional space existed, some, most, or all of what I was being taught was wrong. This realization created a pessimistic attitude I couldn't shake. I became disenchanted with the scientific world. I came to school to find the truth, and now, as I sat in class and looked disappointedly around the room, I became disillusioned. Sadly, I realized many of the people who were here could care less about the truth.

The professor teaching the course was making a paycheck. He was earning a living teaching others what he had been taught. If he was wrong, what the hell, he still got paid.

Most of the students were also a disappointment. They were here because they had to be here. Many of these science courses were required courses they needed to take to complete their college education and get their diplomas. They were merely doing what they had to, to graduate.

As we used to say in the 60s, "It was a downer." The learning technique was also a disappointment. Students would sit in class writing down, without question or complaint, whatever was said. Then they would go home, study it, and memorize it to be able to pass the tests and get through the course.

What was even more painful was the realization I was no better than the rest. I had fallen into the same trap myself. All I seemed to be doing was sitting in class, taking notes, and passing tests.

I also realized any students who had the nerve to stand up and speak out never got the chance. So much knowledge is thrown at students in such a short period of time, there is no time for the "intellectual digestion processes" to take place. There was hardly any time for introspection and contemplation. The mind was never given the opportunity to sift through all of this new information and look for errors.

Inevitably, I became disenchanted with the academic experience.

I also went kind of crazy.

After the experience of the store in Florida, and the transfer to the University in Arcata, I now believed in the concept of "synchronicity". I believed all a person had to do to succeed, was to hold a goal within their mind and wait to be guided into those particular circumstances where the solution would be found. But… I was impatient.

Although I believe in this mental process today, and know it to be absolutely true, at that time in my life I did not know *how* it worked. All I knew then was that impulsive actions seemed to play a greater role in my life than the carefully thought out ones. Impulsive actions seemed to be the actions that guided me to the circumstances where the solution was waiting to be found. However, in attempting to force this process to occur, I ended up going off on a weird odyssey.

Every time I had an impulse to do something, I did it.

One day I thought I would like to go backpacking. Then I thought that if I wanted to go backpacking, perhaps I needed to. Perhaps at the end of the trail I would find the answer I was looking for. So, I purchased a backpack, bought rice, beans, and a few cans of corned beef, and headed down the road. Just headed down the road not having the slightest idea where I was going or where I would end up. Since circumstance eventually led me to a small town in Northern California called Paradise, I thought this was where my next residence was meant to be. Perhaps this was where the answer I was looking for awaited me? So, after returning for the final few weeks of the semester at Humboldt, I quit school and went to "Paradise."

Because of my religious investigations, I was enchanted with the name "Paradise". An enchantment that soon ended when to my dismay, I learned the town was originally called "Pair of Dice". A name given to it because of all the riotous gambling which took place there during the 19th Century Gold Rush. So as suddenly as I became interested, I lost interest in that name. However, because this town was located near another town called Chico, which just happened to have another branch of the California State University system located there, I thought this must have been the reason why I ended up in Paradise.

So, I enrolled. I changed my major to Civil Engineering and spent a semester there.

But if I was disenchanted with school before, I was really disenchanted now. I did not know Chico was notorious for being a "party" school where fraternity and sorority groups dominate the town. It was disgusting to see these young adults seeking to fit in with the expectations of their peer groups rather than seeking to develop their own individualities. They were like plastic people being pressed into look-alike molds. Disappointed, I felt another change was needed – this time in me.

Reluctantly, I gave up looking for the hidden meanings of life in every move I made. Dejected, I left Chico and at the end of another long odyssey, finally ended up going back to California State University at Long Beach. But even though I tried to again force myself to study with the same intensity I did when I was there before, it was no good.

The past eventually caught up with me again. Having to take Third Semester Physics as part of my Civil Engineering requirements, I rebelled. Third Semester Physics deals mainly with Einstein's vision of the universe and his Theory of Relativity. Two subjects I knew for sure had to contain major errors.

For weeks I sat in class, discontented. The evidence for the acceptance of the Theory was very impressive. I reluctantly listened as the professor said; "Einstein's brilliant deduction of the fourth dimension of space-time explains many of the unusual phenomenon of the universe. For example, this fourth dimension of time is responsible for the length shrinkage and time dilation effects which occur at near light velocities." Of course, I did not know then that there is no fourth dimension of "space-time." Even though there is a fourth dimension, a structure within the atom nobody knew existed is responsible for these length shrinkage and time dilation effects: [something I proved in my Doctorial Thesis presented in 2005].

He then went on to say, "Further proof of the Theory of Relativity is found in the observations of very fast moving subatomic particles. The decay of fast moving subatomic particles acts in perfect accordance with the Theory of Relativity. When short lived particles such as the Muon (a heavier version of the electron) are moving faster, they exist for longer periods of time before decaying into smaller particles" [I also did not know then there exists a completely different explanation for this phenomenon and proved it in the PhD thesis.]

He went on to mention another proof of Relativity was seen in an experiment done with an atomic clock. An atomic clock launched into space and orbiting the earth at twenty five thousand miles an hour runs slower than a clock located down here with us. And he concluded by saying, "This slowing down of an atomic clock located within a satellite orbiting the Earth is a further confirmation of Mr. Einstein's famous theory." This statement is totally and completely wrong! This phenomenon is not a confirmation of the Theory of Relativity at all! Later, and again, it will be shown this effect is also being created by unforeseen and unsuspected structures existing within the atom. [Again, this phenomenon is explained in the PhD thesis.]

All I knew in 1978 was that when these observations are added to Einstein's use of Relativity to explain some unusual motions in the orbit of the planet Mercury, I myself had to confess this was some pretty impressive evidence. [But again, as before, this phenomenon too is easily explained in the PhD thesis, and in other scientific papers.] It was extremely hard for me then to deny science's conclusion: "the Theory of Relativity is true, is proven true, and there is nothing more to say." (It was only years later I learned there *is* indeed a lot more to say – much more.)

But in 1978, to say I was discouraged was an understatement. After hearing all of these "seemingly unbreakable" affirmations for which I possessed no alternative explanations, I was deeply troubled. How could all of this evidence possibly be wrong? If only Einstein's fourth dimension of time didn't exist, there would not be a barrier between the three lower dimensions of space and higher dimensions of space. Everything would be fine. But it was ["supposedly"] there, and it seemed to account for many of the strange time dilation phenomenon observed to exist at near light velocities. The evidence seemed overwhelming. And I was worn out.

I didn't believe Einstein was right and I couldn't prove him wrong. I refused to learn what was being taught. And by refusing to learn I couldn't pass tests.

With this rebellious attitude, no matter how hard I tried to fit in and be a part of the academic system, it didn't work. It was only with great difficulty that I picked up the books I no longer cared about and forced myself to study.

Eventually, I knew I had to go. One day, I just got up and walked out. I stood up, right in the middle of class, and with everybody watching me, gathered up my books, walked across the classroom to the door, opened it, and walked out of the academic world forever, [or so I thought].

I must admit I was sorry to go. Not because I was sorry to leave the academic environment, but rather, because I felt I was a complete failure. I had failed to learn where the error in science existed. I wanted to use the principles of science to prove higher dimensional space existed, but I was unable to do it. And lastly, I believed I had failed to acquire the knowledge or the skills necessary to accomplish either of the above goals. Hence I was depressed. [Not knowing then, that I had already acquired all the knowledge and the mathematical skills necessary to prove the Theory of Relativity was wrong. That another revolutionary theory about the construction of the universe was waiting to be discovered by me that would transform science and prove Jesus' location for the Kingdom of God actually exists.]

But the proof had to wait for many years because I was not well.

Considering the last four years of my life to be an utter waste of time, I had no idea what to do. For a while, I just drifted around. In a kind of "purposeless wandering", I drove all over Northern California and Oregon. I had no idea where to go. Life had no meaning. I did odd jobs to make enough money for gas and food and I slept in my car. When I eventually returned to Southern California to search for a more lasting job, I tried to forget about everything.

I wanted to forget about science, higher dimensional space, and get the problem of the location of the Kingdom of God out of my mind forever. I tried to forget about trying to find a way to prove the existence of anything. I was sorry I had ever even heard about the Theory of Relativity, but the rejection never worked.

Eventually I found myself thinking about the problem all over again, and I wasn't able to concentrate upon anything else. Since I never told anybody what I was doing, most of the people who knew me thought I was crazy. Perhaps I was.

Because of this stress, I couldn't stand to be any one place too long. I would get a job, work for a while then quit and move on. I was never satisfied with anything I did. I traveled all over the southwestern United States. But eventually, my wanderings found me back in Florida, where, after a number of years, I passed the eight hour test and became a State Licensed Commercial Swimming Pool Contractor [#CPC 019012]. I built pools or worked as a superintendent for other builders.

Without going into details, suffice it to say, many years passed, and many events occurred. But even though I tried desperately to avoid it, every once in a while I would start thinking about the error in science all over again and my failure to prove the existence of the Kingdom of God. Then, just like a soldier suffering from some sort of Post-Traumatic

Stress Disorder, I would have flashbacks, and suffer terrible feelings of stress and anxiety that sometimes lasted for weeks. While in the interim, I tried my best to cope, to get by and to live with a sense of failure.

And then one afternoon, everything changed. Mediocrity disappeared – exhilaration returned. In a flash of inspiration, I was finally able to deduce the great error which exists in Einstein's Theory of Relativity – the error that allowed me to discover the true vision of the universe; a stunning, spectacular vision no man has ever seen before. A vision of the universe that has allowed me to explain all of the great mysteries of science!

But most important of all, after what seemed like a lifetime of suffering, I was finally able to scientifically prove Jesus' location for the Kingdom of God actually exists. (And an even more shocking revelation about God!)

It happened like this...

PART III

EINSTEIN'S MISTAKE

Chapter 11
I Find the Mistake in the Theory of Relativity…

1989 was a good year. I told myself the stresses of the past were gone, and I was content living a life free of the need to discover the mistake in the scientific vision of the universe. Leave that to others; I was happy in being happy and wanted nothing else. I was making money, and doing the kind of work I enjoyed doing. I had no problems and wanted none.

Then one day the past caught up with me. Everything changed. In one hour my happy though meaningless little existence was blown away. Because of that day, nothing will ever be the same again for any of us. For on that day the great error in Einstein's vision of the universe was deduced.

It all happened on one particularly hot and humid afternoon in Margate, a suburb of Fort Lauderdale. I was working as a swimming pool subcontractor; building homeowner pools for a now defunct company called Swimming Pool World. Working for myself and by myself, I would lay out the shape of the pool, supervise the digging of the hole, form the shape of the pool out of wood, and then finally install the plumbing and rebar.

In the winter, the weather was not bad for this kind of work. But in the summer, it got hot and humid down in those holes. And on this particular day, it was hot, real hot.

It was so hot I could only work for a little while in the hole, then I had to climb out, turn on the hose, douse myself with water, and go sit for a while in the shade of a tree.

Like other construction workers, I had no choice but to work in the intense summer heat of Florida. But to beat the effects of the heat, I had developed the technique of intensely concentrating upon a philosophical problem. I found when I did this the resulting introspection effectively removed my conscious mind from its unpleasant surroundings. These mental gymnastics allowed me to keep working where others couldn't. Although in the past I thought about a variety of subjects, on this particular day, I was thinking about "time".

Time has always held a fascination for me. Many years before when looking for the error in Einstein's vision of the universe, I had researched the origins of time. Or rather, I had tried to research the origins of time. Although there are a number of books written upon the concept of time, they are incomplete. The concept of time is an ancient concept so old its origins are lost in antiquity.

The early civilizations documenting the concept of time were chronicling a concept created much earlier than the creation of writing. (Note: this idea is not speculation. In 1993, it was reported in *Smithsonian Timelines of the Ancient World* , that a carved bone discovered in the Grotto du Tai in France appeared to be a solar calendar dating back to

10,000 BC.)

The important point to remember about the concept of time is that it was developed during an era of great ignorance. The concept of time was developed during a period when man knew nothing about the motions of the Earth, Moon, Planets, or Sun, which is ironic because these motions are, and always have been associated with the passage of time. For example, the day is one rotation of the Earth upon its axis, the month is approximately one orbit of the Moon around the Earth, and the year is one orbit of the Earth around the Sun.

In fact, since all motion is associated with time, on this particular day, I began to wonder about just what motion it was that brought about man's first perception of this most enduring of all "Scientific Ideas"?

Naturally, since both the "day" and the "year" are two important lengths of time associated with celestial motions, I began by telling myself the first perception of time probably began with early man's most rudimentary astronomical observations.

Almost everybody could see day followed night and night followed day. The Sun, Moon, and stars rose above the eastern horizon and set below the western horizon. This entire phenomenon was visible for everyone to observe. However, a very clever individual would have begun to notice an unwavering regularity in these motions and would probably have begun to keep track of them through association with other motions.

Noting the phases of the Moon, he might realize that if he made one mark upon an object for each day of the lunar cycle, he would discover this constantly repeating light show always occurred during the same number of days. He would also observe the night sky, and notice how various star patterns would either disappear or reappear during different seasons of the year. And if he were lucky enough to live in one small geographical location year in and year out, he would notice how the location of the rising Sun continually changes its position each day. Slowly trudging northward for six months, stopping, then heading southward for six months, stopping, and then repeating the whole process over and over again, year after unbroken year. Although this particular phenomenon was probably noticed only by later agrarian societies, which tended to live more stationary lives in contrast to the wanderings of the early hunter, gatherer groups.

I was about to continue on with these mental speculations when I was suddenly halted by an astounding idea I never thought before - was dawn age man really sophisticated enough to recognize such an abstract concept as time?

The concept of time is a very abstract idea. It cannot be perceived with the senses. It can only be deduced through observations associated with motion. This observation is extremely important, for many of our conclusions are based upon our perceptions, but sometimes our perceptions are based upon our conceptions.

Consequently, was dawn age man cognizant enough of his mental processes to differentiate between the two? Is anybody? Spoken more bluntly, did he perceive time, or did he conceive time?

When considering this hypothesis, we tend to forget the concept of time is not a concept we as 20th century men developed. It is a mental legacy inherited from early man and passed on from generation to generation. A concept developed within the reasoning process of the child like superstitious minds of our ancient ancestors - men totally ignorant of the astronomical knowledge we take for granted today.

Today, it is hard for many of us to accept the fact that grammar school children know more about the motions of the Earth and the solar system than the greatest philosophers in the Golden age of Greece. It is hard to believe that six and seven-year-old kids know more about the orbits of the planets than Plato, Socrates, or Aristotle. But what is really astounding, is that these men knew far more than their ancient forbears did who were responsible for developing the concept of time.

Since the workings of the solar system are so obvious to us, it is hard for us to envision a time when men knew nothing about the motions of the Earth. It is also easy for us to forget today that ancient peoples did not know the Earth was rotating - creating the effect that everything was circling it. Or that the changing location of the Earth in its orbit around the Sun was responsible for making different star constellations appear at different seasons of the year; or that during its orbit of the Sun, the Earth's 23.5 degree tilt was creating the four seasons.

In fact, everything ancient peoples were observing in the sky were actually effects being created by motions they were totally unaware of.

It gets even more uncomfortable when we realize our concept of time was developed by men who worshipped fire, feared thunder and lightning as the wrath of the gods, and made animal sacrifices for good crops and successful hunts.

In all honesty, I must admit I was surprised to find myself even thinking these thoughts I had never thought of before. I found myself wondering if anyone else had ever thought like this before? The concept of time is so much a part of our daily lives that to question its validity, even within the privacy of our own thoughts, seems ludicrous – if not insanity itself.

But then, "just because" everyone "knows" time exists, doesn't mean it does. Once upon a time everyone "knew" the world was flat, everyone "knew" the Earth was at the center of the universe, and everyone "knew" Aristotle's vision of an unchanging universe was absolutely correct. Once, to be a member of the educated elite in Europe, everyone studied Aristotle and knew the stars were in stationary positions in the night sky. The so-called "educated people" "knew" the positions of the stars were eternal and unchanging. But what is most disconcerting of all, is all of these totally false and erroneous ideas endured from one generation to another for almost two thousand years.

Man has grown comfortable with his belief in time. But suddenly I wasn't. I began to suspect the concept of time might be nothing less than a total and complete mistake, perhaps the greatest blunder mankind has ever made.

Although the concept of time might have started out as a simple concept, in the 20th Century it has become extremely abstract. A situation which causes it to be suspect. I have learned from my analysis of previous erroneous ideas, such as ancient man's belief in the planets orbiting the Earth upon giant wheels, that the more complex the idea becomes with age, the more likely it is to be wrong. This situation occurs because if an idea is wrong, it constantly has to be amended and re-amended as new information is deduced or discovered.

Startled to be even considering the possibility that the idea of time might be a mistake, I began to wonder if anyone in his right mind had ever dared to challenge this most basic and fundamental of all-human beliefs? Is it insanity to think time doesn't exist?

But if time doesn't exist, what does? What were ancient peoples really observing, the effects of time, or the effects of motion? Although motion is associated with time, maybe this idea is wrong, maybe it is the other way around - *maybe it is time that is associated with motion*!

I was about to go on, but it was just too hot.

I overstayed my time limit in the pool, got overheated, and just barely made it up the ladder. I crawled out of the big hole, and half staggered into the welcomed shade of a Banyan tree. The alarmed homeowner, who witnessed my shaky efforts from his window, opened the door and came outside with a glass of water. He asked me if I was all right, and when I didn't answer immediately, he threw it right in my face; apologizing as he did it because he thought I was going to pass out. Which wasn't true at all, I was merely trying to re-focus my thinking onto the problem of time and didn't want to interrupt my train of thought by talking to him.

When he told me he was going down to the store to buy me some Gatorade, I mumbled to him to go. I was glad to see him go. I was glad he was getting out of the way, because after 15 years of aimless searching, I was finally getting someplace. After he left, I observed clouds coming towards me and knew they would bring a welcome drop in the temperature. Knowing I could return to the pool later and do more work when it was cool, I returned to my thoughts; or rather to one single thought which was uppermost in my mind: if "time" was not creating the phenomenon of time, then what was?

It didn't take long to realize there was only one possibility – *motion*!

Of course! Suddenly it was all so simple. What was early man really observing? Motion! Harmonic motion! And then the truth about the development of time was finally made clear. The development of the concept of time was a mistake. A mistaken attempt made by men ignorant of the motions of the Earth in their efforts to assign a cause to the celestial phenomenon they were witnessing.

I was thunderstruck. The problem with man's vision of the universe was "time". Time itself! "Time" does not exist and never has. The great mistake in Einstein's vision of the universe was suddenly apparent. One of the five fundamental principles of the universe - matter, space, time, energy, and the forces of nature - didn't exist! And Albert Einstein, just like all of his predecessors, had also based his scientific vision of the universe upon something that doesn't exist.

How simple. How extraordinarily simple! I was flabbergasted.

This was one of the great days in my life. On that day, like the flash of light that precedes the explosion, the old vision of the universe was suddenly blown away. The race was on again. The error in mankind's scientific vision of the universe was no longer a possibility, it was a reality. And I was excited.

The universe was not constructed the way we were trained to believe. If I could find out how it was constructed, and if higher dimensional space was part of this new reality, it would prove Jesus knew more about the construction of the universe than all of the greatest scientists who had ever lived. An incomparable discovery of unparalleled importance. A discovery unlike any other ever made in all of human history.

Thinking the above thoughts, I felt a deep sense of responsibility to everyone everywhere. I was in one of the most unique positions experienced by only a few people in all of human history: I and I alone possessed knowledge capable of changing the lives and the destiny of the human race. Knowledge that would directly affect the science, the technology, and the religion of all future generations to come.

In honor of the position fate had bestowed upon me, I quit my job and went to work.

Chapter 12:
The Curious Relationship Between Time And Motion…

In the following months, I spent most of my evenings in the tall, stately library in downtown Fort Lauderdale. It was great to be searching for the truth again, and it felt even greater to have the zest for life return. A feeling I hadn't felt for years. But it had been a long time since I had done any kind of academic study, and I wondered if I was up to the challenge. However, I soon found that old habits die hard, and oddly enough, the construction of this magnificent library helped.

The Fort Lauderdale Library is much more than a library; it is a celebration of libraries. If architecture were music, this place would be a symphony.

I don't recall when I first walked into this library, but I do remember being surprised when I did. I was surprised because instead of seeing the many racks of books I expected to find – a huge gallery five stories high stood silently in front of me. Although there were elevators, one lone staircase somehow found its way up along the western side of this massive and curious void. I use the words "curious void" because every time I climbed this stairway within this building of knowledge, I had the strange sensation death stalked beside me.

I felt this sensation because the stairs were merely slats covered with carpet, and when I looked between them as I climbed upward [especially between the fifth and sixth floors], the lobby seemed dangerously far below. Fascinated by this sight, I realized I was merely stepping upon boards suspended high in the air, and when I did, I knew my mind was wandering and I must quickly re-focus my attention upon the next step. If not, I felt I might lose my perspective, lose my balance, and with just one slip, tumble over the low handrail and spiral to my death far below: a curious feeling within a building whose purpose is devoted to storing the knowledge of the human race.

Perhaps this feeling was intentionally created. Maybe the architect was trying to convey some emotion he was feeling. Perhaps he was trying to tell us all that the empty void represents death; while knowing we are climbing the stairs to obtain knowledge represents life. And if men do not continually focus their concentration upon the next step on the path of knowledge, they might stumble and suffer the ultimate penalty for it. But then again, perhaps I was just daydreaming.

Then again, to dream dreams was why I was there: to dream the dreams no man had ever dreamed before; to think the thoughts no man had ever thought before; and to discover the secrets no man had ever discovered before.

And that is where it happened; in the quiet library standing in the center of the city that had once been the party capital of the United States, I discovered the mysteries of "time".

I would like to be able to say that everything happened dramatically, in one brilliant shining burst of inspiration, but it didn't. Instead it occurred slowly and patiently, as evening after evening I sat before a table upon the sixth floor and looked out through the giant glass windows at the city far below. Many times, I just sat there silently, watching the day end. Watching the shadows lengthening behind tall buildings and creeping up the sides of the ones behind them. Like a silent chorus of motion, millions of shadows all moving together in harmony. An army of effects, all moving in sync to one grand cause - the Sun. [An observation soon to have great significance.]

Within this special library I was happy. The quiet intellectual atmosphere reminded me of college. I studied there and thought there. And it was there, I first began to realize the inextricable relationship existing between time and motion. An inextricable relationship that begins to become apparent the more one investigates the *measurement* of time.

For instance, I asked myself, is it just a coincidence all measurements of time are defined by motion? The answer is no. Everything men can think of that might possibly be used to measure the passage of time is in motion. There is no exception. Even tree rings are laid down by chemical reactions taking place at fixed rates.

Chemical reactions also play a major role in "Biological Time". The DNA in plants and animals responsible for the growth and characteristics of all living things are made of chemicals reacting with other chemicals.

Even when we think we are sitting perfectly still, there are billions of chemical reactions taking place within every cell of our body. Some people think they can detect the passage of time by sitting perfectly still in a dark room. But what these people fail to realize is their body is a microcosm of activity. Their hearts are beating. Blood is flowing to every cell throughout their body. Electricity is flowing from their nerves to their brains. Within the brain, neurons are constantly receiving and passing electric current, generating a succession of thoughts. Which all lead us to the realization that even though we are sitting still, our body is a virtual metropolis of chemical and biological motions.

These chemical and biological motions create a sequence of events, which are responsible for our aging process. The chemistry of our DNA becomes a chemical "clock" that makes our body "tick". Knowing this we also know our ages are not a result of time, but the result of chemical reactions.

The realization that no measurement of time is free of motion inspired my imagination. I tried to think of a way to philosophically prove the existence of this unbreakable bond between motion and time. As a result of these efforts, a fascinating philosophical conjecture was conceived. It started like this, "If all of the motions of everything in the universe slowed down, stopped and then slowly started up again, is there any way to tell for how long they were stopped?"

The answer, after much consideration, is no. Because just as a physical yardstick measures the distance between two objects, a regularly reoccurring sequence of events – harmonic motion – is used as a yardstick of time to measure the distance between a random set of events. However, unlike the physical yardstick, when the harmonic motion used to keep track of time stops, the yardstick of time no longer exists and the distance between random events can no longer be measured. Hence, we are unable to tell if the universe was stopped for a second, a day, a year, or a million years.

Also, in such a slowdown, since the motions of everything in the universe would slow down proportionally to the motions of everything else, we would never notice any change occurring at all. We would never notice anything slowing down, stopping, or starting up again. In fact, nothing would ever appear to have happened at all. A simply astounding conclusion!

Based on this conclusion, I began to understand that time is a function of motion, a phenomenon created by motion, and as such cannot exist separately and apart from motion. Just like the shadows I observed every evening while looking out of the window of the library were an effect, time is also an effect and motion is the cause. (Although later, I discovered motion is not the final cause. Motion itself is an effect being created by still another cause. Time is merely the last effect in a chain of causes and effects. But all this will be explained later.)

After coming to the conclusion time is a function of motion, a phenomenon created by motion, something else began to bother me. Each night when I was about to leave the library, I reluctantly looked at the clock to see what time it was. This nightly observation troubled me greatly. Like the rest of us, I had always associated time with clocks, and clocks with time. I was troubled to think a clock, the most common of all mechanical devices, was keeping track of something that didn't exist.

Or rather, I should say I was troubled until I realized that even though time does not exist, the phenomenon of time does. The phenomenon of time is real. Just like the shadows creeping up the sides of the buildings surrounding the library is a real phenomenon, and just like the motion of the stars apparently circling the Earth is a real phenomenon, the phenomenon of time is also a real phenomenon. It exists as a "real illusion".

Since I was excited about what I had discovered, I made the mistake of telling a friend that time exists only as a phenomenon, and not as a fundamental principle of the universe. For my efforts, I was rudely shown a clock and asked, "If time doesn't exist, what the hell is that keeping track of?"

Because of that little incident, I realized I was playing with fire. I sadly understood anytime you attack a belief, you are instigating an irrational emotional response. Therefore, I decided from then on, I would go within, and tell no one what I was doing. But just in case I was ever confronted with the same response again, I researched clocks.

This research revealed the great fallacy of clocks keeping track of "time" is dispelled when one investigates the role of the clock as a navigational device.

In reality, a clock is a mechanical device keeping track of a fixed position upon the surface of the rotating Earth relative to the fixed position of the Sun when it is directly overhead. The hour, the minute, and the second were arbitrary units assigned to time long ago and are meaningless. Over the years, time has merely become associated with clocks and clocks with time.

When someone asks another, "What time is it?" in reality and unbeknownst to themselves, what they are really asking is, "According to the clock - how far has the Earth rotated since the Sun was directly overhead at noon?" Since the circumference of the Earth at the equator is divided into twenty four sections, each about a thousand miles in width, and the clock is divided into twenty four hours, every hour past noon indicates the Earth has rotated a little over a thousand miles. The further breakdown of the clock into minutes and seconds allows this thousand-mile distance to be divided into thirty-six hundred equal parts, which allows the navigator to accurately determine his position to roughly a third of a mile.

This relationship between the Earthly location of a clock and the position of the Sun is made graphically apparent to modern man by airplane travel. When we travel from America to Europe, we have to change our clocks when we arrive, if not, they will be out of sync with the position of the Sun and their "time" is inaccurate.

The arbitrariness of our increments of time - the hour, minute, second, day, week, month, and year is exposed when considering the rotation of other planets. When men go to Mars, all of these units will have to be changed. Since Mars rotates approximately one half hour longer (24.5 hours) than the Earth's 24 hours, an Earthly clock taken to Mars will soon be out of sync with the appearance of the Sun in the Martian sky. After 12 "Martian days", an Earthly clock taken to Mars will be 6 hours slow. And after 24 Martian days, it will be 12 hours slow. The discrepancy will really become apparent then, because even though the Earthly clock indicates it is midnight, the Sun will be shining high in the sky, showing everyone it is noon.

If a man upon the surface of Mars tries to navigate using Earthly clocks, he will get lost. He will find the Earthly clock is useless upon Mars. To be able to navigate upon Mars, men will have to build completely new clocks corresponding to the slower rotation of the Martian world.

When he builds new clocks, he will find he will have to change the length of the hour, minute, and second to correspond to the longer rotational period of the Martian planet.

If he stays for generations, the earthy concept of the hour, minute, and second will become meaningless. He may eventually find it necessary to substitute the values of the Earthly second with the values of the slower Martian second. A substitution which will affect his values for all the important measurements which are based upon the value of an Earthly second (such as the speed of light). Hence, all of his mathematical equations for objects in motion will also be affected. He will then find it necessary to have two sets of reference books: one for values based upon the measurement of an Earthly second, and one for the Martian second.

Or he may find it necessary to do away with the concept of the hour, minute and second altogether. He may find it more efficient to divide a Martian day into a "metric system" of units, such as tenths, hundredths, and thousandths.

Although this is all speculation, it is a good illustration to demonstrate our concept of hours, minutes, and seconds, are ONLY meaningful upon this planet or upon another planet rotating once every 24 "Earthly hours". When we try to use our clocks upon another planet rotating at a different rate of speed, we find our "earthly clocks" are unsuited for the task. This failure unmasks the truth about the units of time, and we finally begin to realize our concept of the hour, minute, and second is not universal. We begin to realize clocks do not keep track of "time". "Time" is merely associated with clocks, and clocks are merely associated with time.

After writing down these thoughts upon the fallacy of clocks, I decided it was "time" to leave the realm of philosophy and discover a way to prove everything I was thinking. And, lo and behold, after a long search, I rediscovered the Michelson Morley Experiment. [The subject of one of the last lectures I had listened to before quitting school so many years ago.]

Chapter 13
The Michelson Morley Experiment…

At the beginning of this book, I made the statement that it seemed as if some of the major events in my life were meant to happen to compel me to go search for the ultimate mysteries of the universe. And now, while thinking about the Michelson Morley Experiment, it made me realize it was almost as if another impulsive action of mine – that took place years before – was meant to happen to keep me from being misinformed.

Years earlier, when I quit school, I thought my rash act was purely impulsive. I was in Third Semester Physics and had just finished studying the Michelson Morley Experiment, when I got up and walked out of class. But if I had not walked out at that particular point in the curriculum of that particular course, I never would have been able to discover the error in Einstein's vision of the universe.

Like everybody else who studied Physics, eventually, after working on many mathematical problems dealing with Relativity, I probably would have been indoctrinated into believing Einstein's ideas about the fourth dimension of time. Faced with the *supposedly* overwhelming evidence, and the never-ending insistence of today's teachers that Relativity is the truth, the whole truth, and nothing but the truth, it is possible I would have ended up believing Einstein's vision of time was correct. I would never even have thought of questioning the idea of time. Who knows, I might even have accepted the popular belief that anyone who thinks there is something wrong with the Theory of Relativity is either uneducated, or a fool.

Perhaps I might have fallen into the trap that has trapped so many: human arrogance. I might have believed since I had worked so hard to understand the workings of the most difficult theory in all of science, I was to be congratulated, and perhaps there is nothing wrong with it after all. I might have even tried to incorporate it into the vision of the universe I was trying to deduce, and in the end, would have ended up no better off than any of the physicists who dogmatically teach this theory as doctrine. Luckily, I had avoided that fate.

Although at this time in my life I ardently admired Einstein, I never became a believer. A common trait I unknowingly held with a man recognized as the greatest scientist of his era: Nobel Laureate Albert Michelson.

Michelson was a physicist who became famous for creating what has come to be known as the "Michelson Morley Experiment": one of the most important experiments ever conducted in science.

Tragically, this was an experiment most of the people of the world have never heard about. Which is a shame, considering the explanation of this one scientific experiment alone forces us to define precisely how matter, space, and time are constructed – definitions which are then used to explain how the entire physical universe is constructed. Explanations which both directly and indirectly affect the lives of every person upon this planet.

Although it is hard to believe, the explanation of this one relatively unknown experiment was instrumental in creating mankind's present vision of matter, space, and time. Since the physicists of the Twentieth Century also considered this experiment to be a proof of Einstein's Theory of Relativity (never mind that Michelson rejected Einstein's explanation) a little background information on both how and why it was conducted is essential.

It all began towards the end of the 19th century, when two prominent American physicists, Albert Michelson, a Nobel Prize winner, and Edward Morley, his colleague, devised an experiment using mirrors and light to detect the presence of the "Aether wind".

Just as a ship moving through the ocean creates a wake, the scientists of that era believed the earth created a similar wake or a "wind" as it moved through the "Aether" – the name given to the substance they thought space was made of.

To detect the presence of the "Aether wind," Michelson and Morley first built a relatively simple apparatus. This consisted of two sets of mirrors set at equal lengths and arranged perpendicular to each other. Between them was a lightly silvered mirror allowing the light beams to penetrate through it as well as be reflected off of it. The basic idea was to turn on a light source, allow a beam of light to first shine on the silvered mirror where it then split – reflected off both sets of the other mirrors, and was then reassembled at a target.

Because the earth is orbiting the sun at an average speed of approximately sixty-six thousand miles an hour, when mirrors three and four (see Figure 13.1) were parallel to the Earth's direction of motion and one and two were perpendicular to it, the mathematics indicated the travel times between mirrors one and two **should be shorter** than the travel times between mirrors three and four. And the interference pattern created by the different arrival times of these two beams of light at the target should indicate this result, **but it was never seen**. To Michelson and Morley's surprise, the pattern they saw at the target indicated there was **no difference** in the travel times whatsoever?

Figure 13.1 [To simplify the principles involved, the following simplified diagram of Michelson and Morley's apparatus is placed against the backdrop of the Earth.]

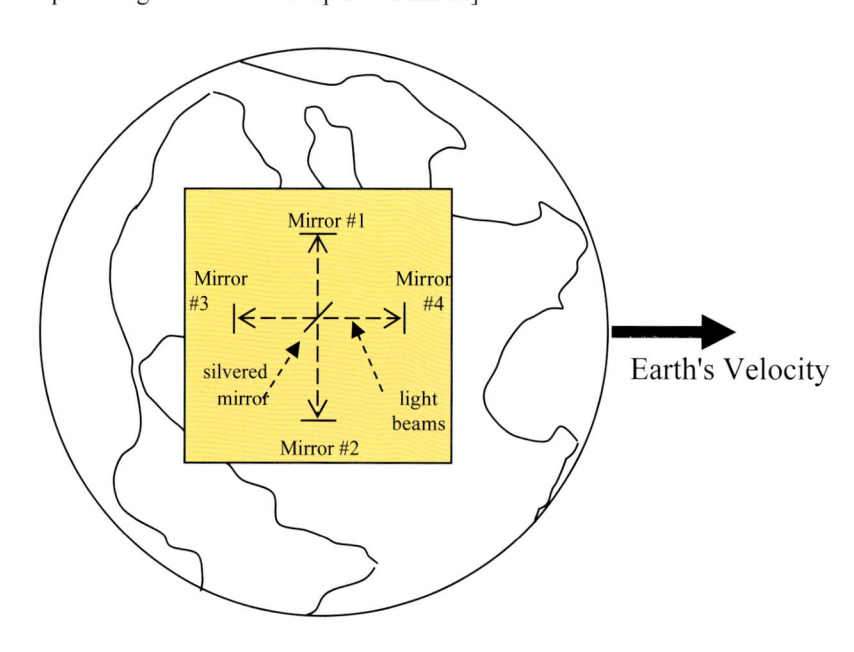

Although surprised, Michelson and Morley were undaunted. They simply built a better apparatus and tried the experiment all over again. But just as before, they still came up with the same results.

Although other people might have quit, Michelson and Morley didn't. They improved the apparatus and tried again; and again and again, but to no avail. The pattern on the screen continued to show the two beams were always traveling at the same speed.

But Michelson and Morley were true believers; they never gave up. They tried for twenty years. They improved their apparatus until it was so sensitive to vibration, it had to be placed upon a two ton slab of sandstone floating in a pool of mercury because an ox pulling a cart down a road close to their building would interfere with it. But even that didn't change the results.

No matter how long they tried, and no matter how sensitive they made their apparatus, the results were always the same – the differences between the travel times they expected to see were not seen. There was no difference between the two travel times whatsoever.

Enter George Fitzgeralds and Hendrik Lorentz.

Fitzgeralds an Irish physicist, and Lorentz a German physicist, both working independently of each other, thought of a way which would make both of these travel times equal. Each one of them proposed that if the Aether wind compressed the matter of the earth in the direction of travel by "just the right amount", **the distance between mirrors #3 and #4 would shorten**. This shorter distance would then **shorten the travel time** of the light beam moving between these two mirrors, making its travel time **equal to the travel time** of the beam moving between mirrors #1 and #2.

Note: in the drawing below, the shrinkage in the diameter of the earth is exaggerated for illustration purposes.

Figure 13.2

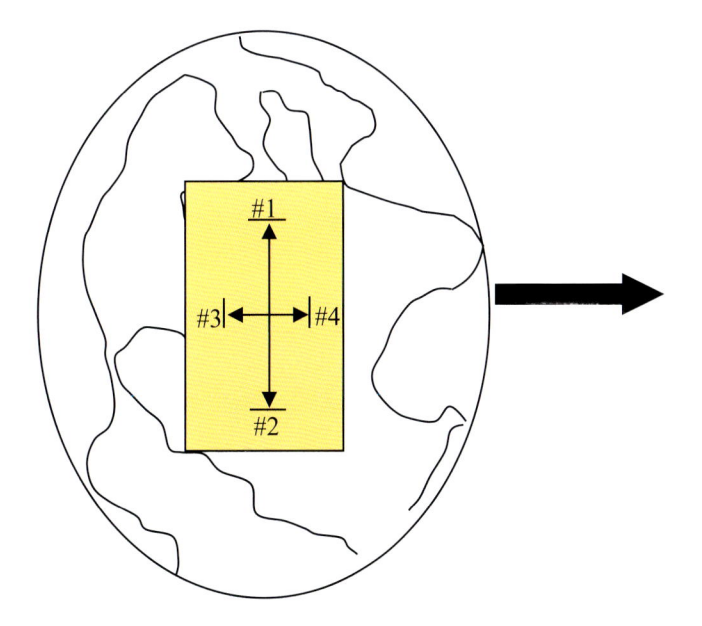

Also, since the Earth speeds up slightly and slows down slightly during the different seasons as it orbits the Sun - although these changes are incredibly minute - because Michelson and Morley's apparatus was now so super sensitive, this change should have been seen too. But it wasn't.

Because it wasn't seen, Lorentz further suggested that **if time "itself" somehow slowed down** by "just the right amount" **no change would ever be noticed**. [Note: Lorentz proposed time dilation before Einstein!] However, the seemingly improbable idea that the Aether wind somehow compressed matter "just enough" to fool the instruments, or that time "slowed down" at high velocities was seen by other rude physicists as an "ad hoc" solution to the problem and was dismissed as being too contrived.

Since no other explanations were available, no other explanations were presented. And the world of physics had to wait impatiently for a solution.

But they didn't have to wait long. When Albert Einstein's ideas began to influence the world of physics, the solution to the problem seemed apparent. According to Einstein's vision of the universe matter was made of something, space was made on nothing, and time was relative and existed as a fourth dimension of the universe he called "space-time." This meant there was no Aether and no Aether wind. This also meant the measurement of time was relative to the speed of the observer. And when these ideas were combined with his brilliant deduction that the measurement of the speed of light was the same for all observers, scientists finally believed they had finally found the explanation for the Michelson Morley experiment.

However, even though Einstein's vision of matter, space, and time has been extremely successful in some areas, it has utterly failed in others. The most important of these failures is its inability to unite the four forces of nature into one "grand unification theory", or adequately explain why they cannot be united: a seemingly trivial technical observation that acquires a status of great philosophical importance. Even those who are uninitiated into the workings of science can easily understand that the true vision of the universe must be able to explain not just some of the phenomenon of nature, *but all of it. Every last bit of it!*

A deduction Einstein himself must have been acutely aware of because he spent most of the remaining years of his life trying to unite the forces of nature into his own personal vision of matter, space, and time. Sadly, failing in the attempt because there was no chance of succeeding. No chance of succeeding because Einstein's failure was not due to a lack of ability or effort, but rather due to his erroneous vision of one of the basic building blocks with which he used to construct *his* vision of the universe – "time."

Because if time is not a fundamental principle of the universe; if time is a function of motion and cannot exist apart from motion, then time is *NOT* a building block of the universe, and cannot be used as one. If time is a function of motion, it is a phenomenon created by motion. Hence, it cannot be used as the fundamental cause to explain time dilation. If it cannot, then the length shrinkage and time dilation effects proposed by Lorentz in response to the results of the Michelson Morley Experiment *are being created by the increased velocity of the atoms themselves as they move through space.*

Consequently, there must exist a hidden, intrinsic relationship between matter, space, and velocity that nobody previously suspected. A secret relationship which unites rather than separates these three seemingly different aspects of the universe's construction. A relationship whose interactions create an exact mechanical and mathematical explanation for the results of the Michelson Morley Experiment - a relationship revealing the *True Vision of the Universe.*

I was excited. Although science presently believes the true vision of the universe was discovered when Einstein introduced the Theory of Relativity, I now knew this was wrong.

The true vision of the universe had not been discovered at all. It was still there, waiting to be found. A sight no man had ever seen before.

Suddenly, I felt like an explorer of old. An adventurer who hears a legend about a lost continent at the end of the world; who gets a ship and sails towards the horizon in search of whatever awaits him.

But this time, it would be no ordinary quest. This would be the ultimate quest for the ultimate prize. The ultimate search for the ultimate vision of everything. And like the view of the Grand Canyon, or Niagara Falls, I knew the ultimate vision of the universe must be an awesome spectacle to behold.

But to be able to be the first man to ever see this true vision of the universe, I realized I and I alone would have to first discover the secret of how matter, space and velocity are related.

Why me? There was no one else. Those who can discover it won't, and the rest never will.

Those scientists who are capable of discovering the answer are unwilling to challenge the Theory of Relativity. They are either true believers in Relativity, and don't think there is anything wrong with it, or they are afraid of the loss of their credibility if they make any controversial statements to their colleagues. So, it is sad to realize those who can discover the truth about the universe - won't.

Unfortunately, the rest of the people in the world will never discover the truth either. They do not have the unrelenting motivation, the "time", or the desperate need to discover the answer to this problem. Therefore, I knew it was up to me.

To me? Suddenly, I felt alone. Who in the hell was I? I was not a physicist.

Feeling inept, I desperately wanted an ally, a seeker like myself. Someone who passionately sought the truth and would go anywhere and do anything necessary to discover it. But I knew no help would be forthcoming. I was alone. It was a hard truth to accept, but after I had reluctantly accepted it. What the hell.

I had started alone, and I must now finish alone. I had come much further than I had ever expected. A little bit further wouldn't hurt. The hardest part of any journey is the last part. So, I prepared myself for the finale.

I bought books on math and physics. I reviewed calculus and differential equations. I went to the library and studied the latest discoveries in science. To increase my self-confidence, I lifted weights in the evening, and did push-ups and sit-ups in the morning. I didn't crawl out of bed; I leaped out of bed. I fasted, lost weight, and felt good. To be a winner, you must first feel like a winner. To succeed, you must already be a success.

And I was successful beyond my wildest dreams.

I took a mental journey through the innermost foundations of the physical universe. I saw what no man has ever seen before, and I discovered the answers to the greatest mysteries of both science and religion…

Chapter 14
The International Association for New Science…

It is always helpful to have a few dollars saved up in the sock. Because I had $10,000 saved, I was able to live on this money for almost two years: carefully rationing my food [a lot of rice and beans], doing only a few occasional contracting jobs on the side, and it worked. There were four mathematics problems I had to solve and working on them 24/7 I solved two of them.

Although it took over a year to solve these first two problems, I was happy with my success [it took another 6 years to solve the last two, but then I was contracting full time]. But before I had to take a full time contracting job and go back to work again in the pool business, I decided to go around to all of the universities in South Florida to show them what I had discovered about the Michelson Morley Experiment and Einstein's mistaken supposition that his fourth dimension called space-time was causing the length shrinkage and time dilation effects at near light velocities.

To my dismay, nobody cared! I went to the physics departments of all the major universities in South Florida trying to convince someone, anyone to look at the mathematical explanation for the Michelson Morley Experiment and nobody would do it! Nobody wanted to even look at the mathematics – or any mathematics – that revealed Albert Einstein had made a mistake in the Theory of Relativity! "Such a thing is impossible," I was told by everyone. "The Theory of Relativity has been proven "time and time" again" or, "You seem to have a serious personal problem if you don't know what a clock is keeping track of?" [Many do not know that a clock is a navigational device originally created to determine longitude for ships at sea and does not keep track of "time!"]

I was also asked by professors, [who never even opened the folder I gave them], "What scientific journal has this mathematics been published in?" or, "What PhD thesis was this mathematics presented in?" Questions that were asked just to make me leave. It was very discouraging. But I was not about to give up for any reason. I next thought about getting the mathematics published in some popular science oriented magazine. I contacted the most popular: Scientific America, Nature, Discovery, and Omni just to name a few. Nobody contacted me back. They did not want to publish mathematics.

Somewhere in this process, a friend told me about a new forum that had just been started by a Dr. Maurice Albertson and his associates called "The International Association for New Science: the IANS". Dr. Albertson's objective was to allow a forum for scientists just like me who could not get their papers published in traditional scientific journals. This appeared to be just what I was looking for! It sounded great!

I immediately contacted the IANS and then sent them a copy of my paper and they contacted me back, saying, "They were impressed"! I was ecstatic! They told me I had been accepted to speak at their new conference that was going to be held in a few weeks on the 17-20th of September (1992), in Fort Collins, Colorado; and that my paper would be published with the rest of the papers from the conference. They would all be put into a book, copies of which would be sent out to all the major libraries in the United States! Wow, suddenly I was home free, what could possibly go wrong? Could anything go wrong? [Yes it could, but I was too excited to even consider any negative thoughts.]

In the next few weeks, I practiced my speech before a group of about 70 construction workers working for me in Stuart Florida; brought in a friend to do my job while I was gone; bought a new suit; got the plane tickets; got a rental car after landing in Denver; and eventually ended up in Fort Collins two days before the conference. But after checking into my room, I made the mistake of going down to the lobby and looking at some of the information available on the conference. Here, I almost fainted!

Many of the scheduled talks were on subjects I call at best, extreme-pseudo-science such as: spoon bending (?), communicating mentally with extraterrestrials (?), picture dousing to eliminate demon possession of the person in the picture (?), Crop Circles (?), Crystal wearing (?), and perhaps the most outrageous of them all: Fire Walking (???)! Yes, actual fire walking on hot coals! They were actually going to hold a fire walking class at night on the lawn outside the hotel!

Regretfully, I now realized I had spent several thousand dollars, and lost additional money being away from work and for what? Nevertheless, on the 17th, I regretfully and remorsefully gave my speech to about 50 people [I had been told I was going to speak to hundreds]. It was what we called in the 60's "A Downer!" But apparently, not all was for naught! For among the 50 was one very important professor from Russia, whom I never met; but who nonetheless, in a week was about to report back favorably to his boss in Moscow about my speech. His boss was that Great Russian scientist Dr. Victor Vasiliev, who's later untiring patronage was about to change the rest of my life forever!!!

Of course, I did not know any of this. I only remember that after giving my speech, I just wandered the hallways for a while wondering what the hell I was doing there. Then, for no reason at all, I decided to go into another of the lecture halls and sat in on a speech already in progress. A woman was bitching to the uncomfortable audience about Jehovah Witnesses coming to her house and telling her kids that if they didn't accept Jehovah right then and there, they were going to Hell! To which she - when she heard about it - instantly reacted by telling her kids while apparently shaking her fists in their faces and yelling [as she was doing right now to the audience], that if the Witnesses ever came back again, to yell at them and tell them: "THIS IS HELL!…THIS IS HELL!" Wherein I and about half of the audience jumped up and got the "Hell out of there"! [And get out fast we did.] It was more like a stampede rushing the double doors, with everybody trying to leave that room at once while she just stood silently on the stage and watched in disbelief!

Out in the hallway, while trying to calm down, I was thinking about going to the spoon bending lecture, but after this episode I said, "to HELL with it," and "THIS IS HELL"!

Mentally exhausted, I went up to my room, turned on the TV, and without even watching it, just kind of passed out in a big chair until dark. When I finally forced myself to get up and go out to get some food [at a McDonalds I think], and while driving back and eating a burger; I suddenly stopped in the middle of the road and watched in shock at about a hundred people standing in a long line upon the lawn in front of the hotel; all with their shoes off, while one at a time, the person in charge would point at the next person in line who would then suddenly take off running barefoot across a bed of red hot coals about 20 feet long and four feet wide placed upon the smoldering burnt out grass of the lawn underneath.

That was it. I spent all my money, $10,000 dollars, and two years of my life for this!? I don't think so! I immediately called the airport to see if I could get my ticket changed to leave early. When the voice on the phone said yes, I got my bag, suit, and left right then. Being a contractor all I could think about while driving back to Denver that night was how they were going to have to dig up and replace the burnt out portion of the lawn?

When I reached the airport, I bought a Denver newspaper to kill time and unfortunately discovered a disheartening article written about the conference entitled, "*Spoon Benders bid for Respectability*." After reading this article, I knew from the disparaging and mocking attitude of the reporter that the conference would never be considered important by serious researchers.

Later, back in Florida, I did not know how to answer when asked by friends "How did it go?" All I could think of saying was, "Well, you kind of had to have been there!" and left it at that. Later, after drinking a twelve pack of beer for solace [and I never drink], I was remorsefully asking myself the same question, "Why didn't I research the conference beforehand?" [Not knowing then, that if I did research it, most likely, I would not have gone; and if I didn't go, just like many other impulsive incidents that ended up changing my life forever, what was just about to happen would never have happened!!!]

PART IV

POWERFUL ALLIES COME TO MY AID…

Chapter 15
A Russian Scientist Was Sent to the Conference in Fort Collins!

I think it was about 6 weeks later that I got a letter from Russia. It was from a Dr. Victor Vasiliev the Chairman of the Electro-Physics department at the *Lenin All-Russian Electro-Technical Institute* in Moscow [the MIT of Russia]. It seems he was intrigued by the strange combination of simplicity and shocking implications regarding Einstein's Theory of Relativity in the speech I made at the Conference in Denver.

I later found out that the Russian government had given him the responsibility for sending out hundreds of Russian scientists to all International Scientific Conferences throughout the world. Scientists whose jobs were to evaluate and report back to him any new scientific discoveries they heard about. It also seems that the incognito Russian scientist at my talk in Fort Collins had greatly interested Dr. Vasiliev in my work. Victor [I later learned] didn't like Einstein because he had learned that it was actually Einstein's wife Mileva who was the brilliant mathematician behind his Nobel Prize winning discovery, and never received any of the accolades Einstein received for her work. [He told me, that this is the reason why Albert gave her all of the money that came with the Nobel Prize. In effect, buying her off!]

Victor later sent me a letter that contained a quiz [in the box below] he used to give to all of his new graduate students. He sent it to me so I could learn why he was not offended by my statement that the mathematical proof of the Vortex Theory reveals that the Theory of Relativity is obsolete. [I had told him that many professors I talked to were offended to hear that Relativity was discovered to be a mistake.] This quiz is more of a rhetorical quiz because the answers to the questions were already assumed to all be given as "Einstein"…

"Unlike Americans, many Eastern scientists do not consider Einstein to be a great intellect, nor anything like the genius Western countries make him out to be. But rather a liar, a plagiarist, and an incestuous adulterer! For example:

Question #1: Who discovered the equation $E = mc^2$?

Einstein?

Wrong!

This equation was first discovered by the great French Physicist Henri Poincaré; and was published in a French scientific journal 2 years before Einstein published his!

Question #2: Who first postulated that space was bent around massive objects such as the sun?

 Einstein?

 Wrong again!

It was done by Paul Gerber (1898), in his paper titled: Die raümliche und zeitliche Ausbreitung der Gravitation; Zeitschrift für Mathematik und Physik, 43, 93. Note: in 1898, this high school teacher from Pomerania explained how light could be bent around massive objects such as the sun <u>many years before Einstein</u>. However, since he had no PhD, his theory was ignored!

Question #3: Who developed the formula for the strange progression of the perihelion in the large elliptical orbit of Mercury around the Sun that Mr. Einstein is given credit for?

 Einstein?

 Wrong again!

Again, it was by Paul Gerber in 1902; <u>13 years before Einstein did it in 1915</u>! When asked about Gerber's mathematics, Einstein acted dumb and said he knew nothing about Gerber's formula; but then later, Einstein tried to belittle him, by saying Gerber's work was nothing! [So how could he criticize Gerber's work if he didn't know anything about it?] Note: both formulas are *identical*; Einstein just changed Gerber's symbols for Eccentricity, Semi-major axis, & Orbital Period!

Question #4: Who first postulated theories about the dynamics of a fourth dimension?

 Einstein?

 Again wrong!

 The originator was Hermann Minkovsky, a German mathematician, professor at Königsberg, Zürich, and Göttingen; who Einstein knew well…because *<u>Einstein was his student</u>*!

Question #5: Who developed the mathematics for his Nobel Prize winning discovery: *The Photoelectric Effect*?

 Einstein?

 Also wrong.

 It was his first wife Mileva Marić! Letters reveal they were partners in these discoveries with her doing the mathematics. Also, the authors for the Photoelectric Effect when originally applying for the Nobel Prize were listed as *<u>Einstein-Marić</u>!* But then later, when they won, Einstein had her name dropped! Because he gave her no credit for the mathematics for the Photoelectric effect, and took all the fame for himself, is probably one of the reasons why she divorced him; and also, for the adulterous, incestuous relationship she found out he was having with his own cousin! So to pay for her silence, and keep his reputation from being destroyed, might just be the reason why he gave her all the money that comes with the Nobel Prize!

Consequently, Dr. Vasiliev's quiz revealed most adamantly he was not a believer in Relativity. All his life he wanted to find an alternative explanation for Einstein's *"postulates"* about the fourth dimension [of which he had no proof!]. This is why he became so excited and intrigued with the ideas of the Vortex Theory: ideas that were in direct conflict with Mr. Einstein's. Victor already was a believer that space was made of something; and not in the *never proven proposal* by Einstein that space was made of nothing. Nor did he want to believe in Einstein's third, even more outrageous proposal: that time existed as a fourth dimension called "Space-time." And the equally disconcerting fact, that Einstein *could NOT explain how Space-time created time dilation or length contraction!* According to Victor, when asked, Einstein just shrugged his shoulders and said it did!

So, Victor was all fired up! When he contacted me, he said his purpose was to get a complete copy of the abbreviated paper that was presented at and published by the conference in their book of papers; and he wanted to see all of the mathematics on the subject for himself.

If it had not been for the fall of the Russian empire in 1986, I would never have written him back. Having been in the Military, Russia was still considered by me to be the enemy. But since America and Russia now had a new and more favorable relationship, and since the cold war was supposedly over, I decided to send him a copy. He reacted favorably, and we corresponded back and forth for the next few months: [including the text of the quiz just presented]. I was heartened by his genuine interest in my work. Because my work was not a national security issue, I was glad I finally found someone in the scientific community who cared. Then I received a desperate telegram from him telling me he needed my immediate help.

It seemed his father was just diagnosed with cancer and immediately needed cancer medicine that Victor could not get in Russia. Was there some way I could get it for him? I telegraphed him back and said "No Problem! I will get it for you!" However, my affirmation was premature. When I tried to get the medicine, I was told by local doctors they could not give it to me because I did not have a prescription. Luckily, Victor was dealing with exactly the right person.

When not building or re-modeling pools for myself, I would hire out as a superintendent and run big projects for other companies. Such was the case now. Presently I was working for the biggest commercial pool company in South Florida called Palm Bay Pools. They made me Vice-president and my main job was to solve problems that inevitably arise during pool construction. So, solving problems was my bread and butter.

Applying my mind to the problem of getting the medicine, I realized that the Russian Embassy in Washington D.C. had to have an in-house doctor because the Embassy would not trust sending their personnel to local doctors. Not only was security involved, but how could personnel who only spoke Russian, communicate their problem to a doctor who only spoke English?

Therefore, I called the Embassy, asked for the doctor, and sure enough, was soon talking to him. I told him what the problem was and after very patiently listening to me said he would be happy to help. That he would call me back later after he called Russia to verify just what and how much medicine was needed.

A couple of hours later, he called back, said if I would buy the medicine that cost $600 dollars, he would send it to Moscow that night in a diplomatic bag. I immediately wired the money to the Embassy via Western Union, and the medicine was bought and sent just like the good doctor said. The next day the doctor called me back and said the Russian Ambassador would like to meet me. When I asked why, he said that the whole embassy was aglow with the knowledge an American citizen was willing to help a Russian citizen and spend a lot of money [in Russia 600 dollars was a lot of money: three months' pay!] for no other reason than that of compassion for another human being. [The doctor actually had to choke back tears as he was speaking.]

I was stunned! I did not know what to say. When I composed myself I told him that - that is what we do in America. That I was honored to be able to help. I also told him that I was just an ordinary Joe. Nobody special! I thanked him for his invitation to meet the Ambassador but would not know what to say; and that I was in the midst of doing a big job and it would be difficult for me to take time off. [This was just an excuse, because at that time in my life, I would have felt out of place formally meeting an important ambassador]; but I could not get away from Victor.

Victor wrote and told me that I had saved his father's life, and he would never forget it! That I was his friend for life, and he was going to get my discovery about time not existing published in Russian Journals.

I tried to tell him he owed me nothing, it was the Good Lord's doing, but to no avail. Victor was on a mission, and nothing would stop him. During the next year he translated the paper into Russian and sent it to friends of his in universities all over Moscow. Five PhD's in physics, four of which were department heads, wrote Peer Review Letters of confirmation on the official numbered stationary of their University's; acknowledging they had read, affirmed, and analyzed the mathematics of my thesis; declaring it "to be true and correct."

For the next two years, Victor sent the paper out to international conferences all over the world. It was also published in the Distinguished Russian Scientific Journal: <u>The Journal of Membrane Science and Technology</u>; then in the following:

http://www.new-philosophy.narod.ru/mm.htm (in Russian)

http://www.new-philosophy.narod.ru/RGM-VVV-RU.htm (in Russian)

http://www.new-philosophy.narod.ru/MV-2003.htm (in English)

http://www.new-idea.narod.ru/ivte.htm (in English)

http://www.new-idea.narod.ru/ivtr.htm (in Russian)

Although all of this effort by Victor was great, nothing came of it. Those who believed in Relativity continued to believe in it, and since I had no experiment able to confirm this Vortex Theory with, it was slowly fading into obscurity.

Friends in America told me I needed to contact the "common man." Their argument was this: "Well schooled lawyers argue a case in court, but the guilt or innocence is eventually decided by a jury of the common people: most of which are totally ignorant of the law." The same for science, you need to write something the everyday Joe & Jane will understand. These are the people to whom your science will eventually affect the most anyway. So I did.

Because I had trouble just writing letters, it was a long drawn out painful process. After months of working on it at night and on weekends, I finished a short book version, made ten copies, and gave it to ten friends to read.

Because they were friends, they later all told me when questioned, "Y'all, it's great. It's terrific. I really enjoyed it, y'all, really." But the look on their faces told the true story; and none of them could tell me what happened after page three!

Sadly, I knew it wasn't very good.

Troubled and disappointed I said, "To hell with it! To hell with it all! Who even cares about the truth? If men choose to be dumb, let them be dumb. I'm finished!"

So, I sat the whole thing aside and forgot about it.

However, like so many times in the past, I did not know then that the synchronistic wheels of destiny were turning again, and a totally unforeseen and unexpected fate was getting ready to put me on the quest again. Her name is Jeannie!

Chapter 16
Fate Brings Me Another Powerful New Ally…Her Name Is Jeannie!

It was on a Sunday in the hot Spring of 1996. Six months had passed since I wrote that terrible book; and I was happy. No longer was I going to be consumed by the quest. I was free of it all and working now – I was going to make some money just for myself – to have some fun and finally somehow enjoy myself. Then my truck broke down.

I was driving east on Sheridan Street in Fort Lauderdale, Florida just before the I-95 Freeway. I was in my old pickup truck, heading back to my little apartment on Hollywood Beach, when a bearing went out on the right front wheel.

Because all of the garages are closed on Sunday, I knew I could not get it repaired by someone else until Monday. Also, if I had it towed to where I lived, it would be very difficult or impossible to work on it in my narrow parking space. So, my next thought was to think about where I could keep it overnight.

Everyone I knew lived on the other side of town, except for one friend, [call her "Dee"] who lived about two miles away; she was one of the people I had given the short book version to. I had known her for many years. Her late husband and I used to be partners in the pool business, and I knew she wouldn't mind helping out. I called her up and she said, "of course it was all right". So, I called a tow truck. This "person" charged me fifty dollars to make the ten minute trip and I was upset.

But Dee told me, "Forget about the truck driver."

She then changed the subject by saying some friends of hers from England were staying there for a day before they fly back tomorrow. That they were out at the moment to get food and would be back soon. So, I got out my tools, jacked up the truck, and started taking off the front wheel when a jeep pulled into the driveway.

A man and a woman got out. I recognized the man [call him Pablo], but the woman I had never seen before. Although she was plainly dressed, there was something very attractive about her. She had a warm and friendly smile and appeared to be about my age. I was later told her name is Jeannie.

After a while, I decided to take a break and went into the backyard where Dee, Jeannie and Pablo were seated at a table; Dee handed me a coke, and I sat down. After some small conversation had taken place between the four of us about the truck, Pablo and Dee got up from the table to move some roofing material stored on her back porch leaving me and Jeannie alone.

Making conservation, I asked Jeannie what she thought of America. She told me it was a wonderful country. She explained how in England Pablo and his wife [call her Janna] were good friends of hers; about how she wanted to attend a healing conference in Tucson, Arizona, but didn't want to go alone; that she enjoyed being in America, thought it was a great country and was sorry she was going home tomorrow. Then she changed the subject and said something odd…

She said she was sorry she was leaving because she had read a wonderful little book about "Time" just before she left England, and would have liked to meet the author before she went home…

She began to describe it, and as she did, it sounded exactly like mine. I couldn't believe it.

Surprising her, (and myself), I suddenly jumped up from the table, went out to my truck and came back with a copy of the book and put it down on the table.

"Was this it?"

"Yes…that's it…that's the one I read!"

"Well………………..I'm the author!"

We just sat there staring at each other in a state of shock!

I would like to have said something to her, but I was confused. Just how did she get a copy of my book and how did it get to England?

But I soon forgot about the book.

The more I looked into her eyes, and the more she looked into mine, it was somehow as if we were old friends from long ago who had never met!

When Dee and Pablo came back to the table, and they both heard the story, Pablo reluctantly confessed he was the one who made a copy of the book and took it back to England.

Six months before, when he was visiting Dee, he had seen the copy of the book on her desk. He had looked at it, found it interesting, and since he couldn't buy one, had secretly made a Xerox copy of it. He had taken it to England and had shown it to his wife and friends, including Jeannie. But she was the only one who had actually read it.

We were all astonished.

Jeannie asked me when I was going to have it published and I told her I was fed up with it. I told her how every scientist in America who was capable of understanding the mathematical proof was ignoring it because I was not a Doctor of Physics, but a construction worker inspired by the words of Jesus in the New Testament. [And was furthermore told by the scientists I talked to in South Florida that religion and science should never be mixed!]

I told her it would probably never be published.

Later in the evening we had a chance to talk again in private.

She insisted it was too important a work to let die. She said I had to get it out to the people of the world because they needed it to overcome the atheistic attitudes dominating present day science. "God needs to be put back into the scientific equations explaining the workings of the universe" she said. "That the very people who think they don't need it, are the very people who need it the most; and just like the "me" of many years ago, they need this knowledge to regain their faith in God too."

Ten months later, we were married.

Our friends said it was beyond incredible, it was Divine Providence! Destiny was set into motion by the invisible hand of the lord. It didn't just happen, it was meant to happen, it was meant to be!

Everyone who has heard the story is amazed at the incredible synchronicity of the events that occurred on that day. And the equally shocking fact that in a distant foreign country [not one next door like Canada], but in one located across three thousand miles of ocean, providence, like some invisible messenger from heaven, caused Pablo to see and copy the book and to give it to Jeannie. I say Jeannie because she was the only person in the world who read all of it! [Pablo admitted he only breezed through it but never read it!] Then, after reading it, Jeannie wanted to meet me, but I was too far away. The reason she thought she came to America was to attend a healing conference. But then, on her last day in America, the real reason why she was here was finally revealed: she wanted to meet the author of the book, and like a miracle taking place before her very eyes, out of nowhere I suddenly appeared!!!

To some atheistic scientists who have since heard the story [and don't want to bother considering the ramifications], it is just casually dismissed as being "just a co-incidence" and nothing more! But to others, it is something much, much more! It has caused great excitement. It was as if we had both come to some sort of divine mystical crossroad where our two lives were destined to intersect, to meet and become one! Was it just a co-incidence or invisible divine intervention? Some have said that our unconscious minds were probably in communication with each other and that is why I was led there. However, what about the wheel bearing; that had nothing to do with our unconscious minds? Was it meant to break on that particular day, in that particular place? How did that happen? Was this also just a co-incidence?

And me, like a soldier of the lord, being called to duty by the powers on high, suddenly, and miraculously appearing to her at that exact moment in her life before she would have left America and be gone forever? It is simply incredible!!!

Friends also keep marveling about the "what ifs": what if I hadn't been driving on that particular road; or what if the truck had broken down at any other place in town; or what if she had not been there at that precise location on the day before she was leaving…we would never have met!

But I have another idea: it was another one of those special preordained events in a long string of preordained events, necessary to happen to be able to first discover, and then make the people of the world aware of this special knowledge that has come to be called *The End of the Concept of Time*: people who would not care about the science, but would care about the shocking discovery that Jesus Christ as the Messiah knew more about the construction of the universe than all the greatest scientists who ever lived, including Mr. Albert Einstein!!!

Jeannie urged me on. She is the one responsible for having *The End of Time* book written and published. I would never have done it if it hadn't been for her.

As I look back at my life, it is almost as if all my life I was an insignificant pawn being slowly pushed secretly across some massive invisible chessboard in a mighty game of chess between science and religion: an unseen pawn hiding at the edge of the board, waiting to suddenly and miraculously appear at the end of the game, surprising the opposition, and checkmating the dark king of atheistic science.

And when I say I was pushed, I mean pushed. When I tried to quit, I was not allowed to quit. Every possible obstacle I encountered was overcome by me or for me. And when I was down and out in the last round of the fight, and thought I was finished, someone always miraculously appeared, reached out and gave me a hand up, and urged me onward: forever onward! It is almost as if God now wants the people of the world to know these ultimate and final truths about the universe and won't settle for anything less.

In Russia, amazing things were happening too!

Chapter 17
The Great Russian Scientist Becomes My Ally...and Mentor!

After Victor's translation of the explanation of the Michelson Morley experiment was presented as a thesis to select members of the Russian scientific community, it was subsequently analyzed, confirmed, and endorsed by five Russian Professors required for PhD Peer Review in Russia: four of whom were Chairmen of their departments in their universities:

> Dr. Alexander E. Filippov, Chairman of Department of Physics and Leading Engineer[*] of the V.I. *Lenin All-Russian Electro Technical Institute* (VIE);
> Dr. Alexander N. Vasil'iev, Chairman of Department and Leading Engineer of the *Moscow Power Energy Institute (Technical – University)* (MEI);
> Dr. Stanislaw I. Gusev, Chief of Department of Electrical Engineering, and leading Engineer of the V.I *Lenin All-Russian Electro Technical Institute* (VIE).
> Prof., Dr. Vladimir N. Puchinskiy, Leading Engineer of the *Moscow State Technical University* (MAMI);
> And of course Prof., Dr. Victor Vasiliev himself, Chairman of the Electro-Physics Department, and Leading Engineer at the *Lenin All-Russian Electro Technical Institute* in Moscow.

Hence the mathematics was now officially accepted as being "Peer Reviewed" by Moscow's scientific elite. So, it was then deemed "fit" for presentation in a new pamphlet called "*The Bases of the Vortex Theory of Space;*" reviewed and certified by the renowned Dr. I. S. Slutskin in November 2002. Then it was published by the house of ZNACK, and sent by Dr. Slutskin to Universities all over Russia [he sold many of the copies and made a few bucks off of it! (Capitalism had finally come to Russia's scientific community!)].

Consequently, because of this pamphlet now circulating among Russian Universities, the mathematics was finally brought to the attention of Dr. Konstantin Gridnev, Chairman of the powerful Nuclear Physics Department at the "Harvard of Russia" – St. Petersburg State University, in Petergof [The world-renowned university founded by Peter the Great in 1724.]

Dr. Gridnev was so impressed he bought a ticket and traveled all the way to Moscow by train just to meet with Victor. Then he got a grant from the Russian Academy of Sciences to come to America and visit me! Imagine that! This distinguished professor who held three PhD's in physics, mathematics, and chemistry; who could read, write, and speak four languages; was a member of the Russian Academy of Sciences; who had written numerous text books on nuclear-physics; who had published more than a hundred scientific papers on nuclear physics and presented them in international conferences all over the world – actually asked for and got money from the Russian Government to come to America and visit a "construction worker!" It was not only unprecedented; it was unheard of !!!

---

[*] Note: in Russia, Leading Engineers are the equivalent of the West's Tenured Professors

But come he did. He spent two weeks at my house, stayed in the guest room, and day after day, we went through all of the mysteries of physics and astronomy the Vortex Theory of Atomic Particles could then explain [about 80 then; it has since expanded and now has explained over 120!]. Konstantin was also a Master Chef, an excellent cook, and knew local recipes from the many different countries all over the world he had visited in his long and successful career as a nuclear physicist. We ate well.

Konstantin should more appropriately be called the "Russian Bear." Not only is he built like a heavyweight Russian weightlifter, he is a loyal son of Russia: always raising his glass and drinking a toast "To the Fatherland." He was in his late 60's when I met him. He had a wife living in Moscow and a son who was teaching physics in Germany. But sadly, most of his family had been murdered by the Nazi's in World War II. As a little boy, horrified he hid in the woods and watched while his father and sisters were forced to stand against the side of their farmhouse then machine-gunned by Nazi Storm Troopers. He buried them himself! Perhaps that is why he became an Atheist.

But Konstantin, more than anything else, was also the consummate scientist. His life revolved around science. He "ate, drank, and slept" every bit of knowledge about science he could find. It was almost as if he was trying to fill some void in his life.

However, what he was most interested in was the mass of the Higgs Boson particle [nobody knew what it was] that CERN was in the process of trying to find in their Large Hadron Collider located on the border between France and Switzerland. It seems another friend of his was the director of CERN and was now looking for the Higgs particle whose mass was the subject of many arguments among scientists. So, the question he asked of me was this: "Were you able to determine, using this new Vortex Theory, the mass of the Higgs Boson?" Without any hesitation I told him "Yes"!

He was suddenly ecstatic and excited…he could hardly wait to ask the next question, "Well then, what is it?" With a carefully assumed and enigmatic expression on my face, I slowly and solemnly told him: "Z E R O"! "Zero? You're telling me Zero?" "Yes, my good friend, "Zero": because it does not exist! Mister Higgs and the entire world of physics is wrong because it is not a "particle" that creates mass, it is the distortion of the space surrounding "the so called subatomic particles of the universe" that creates mass." He suddenly became dead silent, and during the next two hours I explained exactly how both mass and the force of gravity are related to each other and are created by distortions in space itself; and this is why each is, and has to be a part of Newton's famous equation explaining gravity!

I told him, "…It is less dense space, not bent space that creates the force of gravity." Also, if indeed space is made of nothing as Mr. Einstein said, then how can something made of nothing be bent!!! He said nothing so I continued by next explaining Quantum Gravity to him. [He said later that he had always believed in the possibility of quantum gravity but could not explain how or why it existed.] But more important, I explained how the additions of all the Quantum Gravity spatial distortions created around all the protons and neutrons that exist within the interior of a celestial body add up to create the massive gravity effects we observe surrounding planets and stars. Konstantin didn't know what to say. He later told me he had always been a reluctant, sort of quasi-believer in Einstein's explanation that gravity was caused by "Bent Space;" mainly because he had never heard of another viable alternative that successfully contradicted it. So now that he had finally heard one, he no longer knew what to believe! [His incredible mind was in a state of flux!]

Poor Konstantin was suddenly shocked and "burned out" [to use another expression from the 60's]. He had to go and lie down and take a nap for a couple of hours. He told me it was going to take him a while to digest what I said because although he thought he had heard every possible explanation for both mass and gravity, this revolutionary knowledge was something brand new, and it was going to "take him a while" to assimilate it!

After that, Mr. Konstantin was now suddenly "On-Guard" and "Alert," ready for anything and everything. He was no longer surprised or shocked to next hear revolutionary explanations for some of the greatest mysteries of science he had spent his entire life trying to understand: such as "Quantum Entanglement," something a mystified Einstein had called "Spooky Action at a Distance". Next was the destruction in Konstantin's long held belief in the so-called "Dirac Sea of Particles," proposed by Paul Dirac in 1928, and used by Dirac as the bizarre and mistaken location for the new group of particles he was theorizing to exist in the universe: "anti-matter"! He was absolutely brilliant to propose the existence of anti-matter, yet did not understand how it was created, or where, or why it existed. Therefore, he proposed that they existed "under matter"! [What!] In later years when others proposed the existence of quarks, they were also mistakenly theorized to exist as a vast…"Sea of particles existing under the matter (?)" of the universe!

I explained this vast Sea of particles existing "under matter" was ridiculous! There is no up or down in the universe, so how can something exist "under" something else? How can a grownup propose such a ridiculous idea and how can other so-called "educated" grownups believe it?

I told Konstantin that although Dirac was indeed a great theoretical physicist and a brilliant mathematician, he was totally wrong in trying to explain where, how, and why anti-matter existed. I told him that the Vortex Theory now easily explains matter and anti-matter as being the ends of tiny three dimensional vortices that flow into and out of the fourth dimension, just like the two ends of a pipe...

This simple yet elegant idea also explains why both have the same amount of charge, but opposite signs from each other [one negative, the other positive]. The positive charge is caused by space flowing into the hole at one end, while the negative charge is caused by space flowing out of the hole at the other end. It also explains other phenomenon such as the law of the Conservation of Charge, and "Mutual Creation" the fact that in particle collisions, matter such as protons and electrons are always created in matter anti-matter pairs. In summary, there is now a completely new and easy explanation for the existence of anti-matter. The idea that anti-matter "exists beneath matter" in a gigantic sea of particles is out, and the simpler explanation is in: Mister Occam, of "Occam's Razor" fame, would be happy!

After "blowing Konstantin's mind" with that explanation, I then told him that all 20th Century scientists were handicapped in trying to make explanations for the creation of the universe, because they blindly accepted Einstein's erroneous belief that space was made of nothing. This was too much, and Konstantin had to go lie down again. I told him later when he woke up that the mathematics explaining the Michelson Morley Experiment made via the Vortex Theory of Atomic Particles revealed space had to be made of something. Nor was this a return to the old Aether Theory where it was believed that both matter and space were made of the same substance [with matter being made of condensed space: like ice in water → probably where the idea originated]. Instead, according to this revolutionary Vortex Theory of the universe, space was discovered to be made of something and matter [the subatomic particles of matter like protons and electrons] were discovered to be made of nothing at all! These subatomic particles were actually three dimensional holes existing upon the three dimensional surface of a fourth dimensional volume. A volume made of space only, with no "space-time" characteristics!

Consequently, since most scientists believed Einstein's mistaken vision that space was made of nothing, they were forced to try to explain all of the phenomena in the universe with the use of particles: such as the Higgs boson.

The shocking reality was this: "everything that exists in the universe is created out of the substance space is made of and not particles"! What we call particles [protons and electrons] are actually holes in the surface of three dimensional space; while Quarks are holes in higher dimensions of space that exist within the interior of the three dimensional holes! The forces of nature are created out of less dense and flowing space; energy is created out of denser, contracting and expanding regions of space; and time, "Time itself" does not exist: it is a phenomenon created by motion; a "Shadow" of motion! Without motion there is no "time." [Note: even the rings of trees are laid down by chemical reactions taking place at fixed rates.]

Even more shocking, CERN just spent 13 billion dollars to build a machine to look for something that does not exist!!! What a joke. However, it is not a joke for poorer countries such as Italy, and Greece who are paying money to CERN they can ill afford to waste; yet have to pay, because they are locked into an international treaty they have to comply with.

After our first three days of tough head to head revelations and confrontations, Konstantin was hit with so many revolutionary explanations so fast, he was overloaded and needed to take a break. To his credit, when he did not agree with something right off, he was quick with a scolding, "Nit, Nit!" But eventually, he calmed down later after the explanation was presented to him and he became lost in introspection. It was tough on Konstantin; his comfortable world was being torn apart. I felt sorry for him because I had experienced the exact same feelings when I first discovered this revolutionary knowledge that conflicted with everything I had learned at university. [But I had time to digest it over a period of years; while Konstantin was hit with it all at once!]

So we took a break by going to the Florida Keys with my friends Fabian and his wife Rhonda. We went diving on the beautiful tropical reefs around Islamorada. Konstantin had brought his diving mask from Russia that had his prescription ground into its faceplate to correct his poor eyesight. We all had a good laugh later when we went to a restaurant serving different kinds of fresh fish for lunch. They all ordered grouper, swordfish, and crab, but when it came my time to order, I ordered a Cheeseburger and fries; wherein Fabian started singing Jimmy Buffett's famous song: "Cheeseburger in Paradise!" To which all of the restaurant patrons started laughing and clapping.

By the time Konstantin left, we were good friends. He invited me to come to St. Petersburg State University later that year and speak at an international conference he was giving on physics. I was sorry to see him go. He was what I had always imagined a great scientist to be. A man, who when confronted with the possibility that everything he believed about the construction of the universe was wrong, was willing to adapt, to change. It takes a great man to do that.

It was also a shocker to learn he had a photographic memory. I had only heard rumors of such people, mainly in myth. But Konstantin was no myth; he needed no books; knew every formula in physics, mathematics, and chemistry. I was stunned to observe his amazing ability firsthand, and remarked he was like a "walking encyclopedia." To which he joyfully replied that that had been his nick-name at university. It is also interesting to note that this former diehard Atheist later became interested in, and started reading the Bible after learning that the mathematics of this thesis, and all its subsequent revolutionary explanations for all the mysterious phenomenon of the universe were based upon just one quote made by Jesus in the New Testament! Because he now trusted me, he also shared with me some of the intimate details of his life, such as when his mother died; and how she appeared to him in his mind on that day; an experience that directly conflicted with his belief in atheism; but was now explainable via the Vortex Theory.

[I was broken hearted when years later, I heard he had died. But I live in the comforting solace that the science of this Vortex Theory of Atomic Particles inspired him to become a Christian! And I know that the good lord would be happy to welcome this great man into heaven.]

I mention all of the above about Konstantin because he later became my mentor and advocate. Both Konstantin and another great physicist Dr. Lev Ivlev, with PhD's in Geophysics and Aerosol physics [the science of weather]; the author of 18 books on aerosol physics and geophysics with 75 international scientific papers to his credit; who was also the Chairman of the Aerosol Physics Department at St. Petersburg State University, were both responsible for getting me a PhD in Nuclear Physics from the Russian Ministry of Education. This branch of the government is responsible for awarding all higher academic degrees in Russia [this eliminates both fraud (people buying degrees) and nepotism].

Konstantin's last question asked of me before he left to go back to Russia was this: can you devise some sort of experiment to prove what you have told me? If so, we can do the experiment at St. Petersburg State University in Peterhof when you come to the conference?

PART V
I DEFEND THE THESIS...

Chapter 18
Fabian and I Go to Russia...

It took me a while, but I finally succeeded in doing what Konstantin asked. It took six long months to devise and build an experiment, but I designed and built an ultra-sensitive laser interferometer used to change the speed of light! The experiment used a powerful green laser and equally powerful electro-magnets that when turned on, were used to *increase the speed of light*. I chose this experiment because not only was it dramatic, it demonstrated something that should not be possible: increasing the speed of light. This was something Einstein said could not be done; that the speed of light will be measured the same by all observers everywhere in the universe and cannot be changed. The speed of light can be slowed down such as when entering a medium like air or water, but never speeded up. But Mr. Einstein was wrong! [Victor was later ecstatic when he heard this, and even more ecstatic when he saw the experiment succeed with his own eyes!]

My good friend Fabian, a computer genius from Argentina and about 20 years younger than me, enters the narrative here because he felt he was trapped in the mortgage business and although he was making a lot of money, his heart was really in this new science revealed by the Vortex Theory of Atomic Particles. He wanted out of the mortgage business because of his fascination with science, but had to stay because of the money.

When he heard I was invited to the conference in Russia he wanted to go too. So, in the summer of 2005, we said our farewells to our wives, and Fabian and I set off on a journey to Russia. Called the LV National Conference on Nuclear Physics, FRONTIERS IN THE PHYSICS OF NUCLEUS, it took place on June 28-July 1, 2005; and was held in the beautiful city of St. Petersburg.

This beautiful city, built right beside the Gulf of Finland in 1703 by Peter the Great possesses many massive, magnificent government buildings made of expensive marble and topped by large bronze statues of saints and heroic leaders from Russian History. The city also possesses canals similar to Venice; and houses the famous Winter Palace; and the Czar's spectacular Summer Gardens.

These immense gardens stretching for about four kilometers, have within their interior, fairy-tale buildings topped with glistening, gilded minarets adorned with flag poles flying colorful banners; towers of gold, surrounded by enormous fountains shooting jets of water 60 feet into the air. Huge golden statues standing beneath tall trees are everywhere; making it the proud heritage of the people of Russia and their most favorite tourist destination.

When we arrived in St. Petersburg we were met at the airport by Konstantin and Victor, who had traveled by train from his home in Moscow to attend the conference. Unfortunately, we learned that our luggage with the experiment in it got lost at the airport in Amsterdam, and all we had were the clothes on our backs, [and some cheese and bottles of Vodka we had bought as gifts for Victor and Konstantin at the duty free stores of the airport]. But just like before, this also proved to be another one of destiny's, "meant to be" synchronistic events.

In all probability, the experiment would never have cleared the x-ray machine at customs. In one suitcase we had two super powerful .3 Tesla electro-magnets strong enough to rip the fillings out of a person's teeth, or kill a man with a pacemaker; we also had two extremely powerful green lasers with a two mile range and capable of blinding a pilot flying overhead in an airplane.

When we first arrived, we looked at the long line of frustrated people waiting to clear customs by first having their luggage x-rayed by a lone antiquated x-ray machine [a second one was broken]. Because each procedure took several minutes, Fabian and I looked at each other in disappointment, knowing this would be a long wait when our luggage arrived.

The next day after our luggage had arrived, it got lost again. Konstantin and Victor, who came with us, later found the room where the luggage was being stored; and when the door was unlocked and opened, we saw it was filled with hundreds of unclaimed suitcases piled on top of each other in a big, jumbled mess.

After an hour or so of searching and after moving piles and piles of suitcases, we finally found ours. But when we carried them out, another long line of a hundred people were there; standing impatiently, waiting in front of the lone x-ray machine. Konstantin and Victor got fed up. Victor, a tall stately man whose air of importance commands the respect of others, began to angrily tell off the airport security guards standing by the x-ray machine.

At first I thought he was going to get himself arrested; however, after showing these men credentials that identified him [as I later learned] as the former head of the Communist Party in Moscow [a title that was still giving him enormous political influence in Russia] the guards backed off. Konstantin, seeing an opportunity unfold by the distracted guards, suddenly grabbed our bags, and quickly walked past the x-ray machine, and out of the airport.

Fabian and I, not knowing what to do, followed. I looked back for Victor, but he was still giving those men hell at the x-ray machine.

Afterwards, Konstantin told us that if the airport security had seen our laser beams, and powerful magnets, in all likelihood, they would have been confiscated on the spot. However, it didn't happen; because losing our luggage was, what I like to now call: another of those "meant to be synchronistic events." Because without the lasers or the magnets, the experiment would never have taken place; and the subsequent chain of remarkable events about to occur would never have happened.

Chapter 19
We Increase the Speed of Light…

Fabian and I finally arrived and officially checked into the huge but strange looking hotel we were to stay at. We later learned that the hotel now housing all of the attendees of the conference was once – with its three foot thick walls to protect against the winter cold – the former stable built for the Czar's horses!

About one hundred scientists attended. Because Fabian has what is called the "Gift of Gab," which made him a very successful mortgage broker, he soon had met and talked to so many people who spoke English that everyone there seemed to know his name.

After we had settled in for a day to overcome the jet lag, Konstantin took us to the university. The experiment was to be done in one of the huge buildings St. Petersburg State University is known for: a massive structure four stories high and a thousand feet long, housing the renowned V. A. Fock Institute of Physics. We were given a laboratory on the third floor, and Fabian and I went to work. We unpacked the experiment and set it up on several of the massive tables in the lab.

I designed and built the interferometer, but Fabian made it work. I had no patience, but Fabian did. He enjoyed and possessed the untiring concentration & ability to work for the hours it took to set up this supersensitive equipment that even the act of breathing interfered with. This is when we encountered our first problem; there was too much vibration in the building.

When we performed the experiment at my house in Florida, it was done on the solid concrete floor of the house, and performed at night when no cars were on the street. However now, we were on the third floor with people walking above and below us. When I told Konstantin, he immediately emptied out half of the building. Seven hundred professors and students had to go [the students were happy to leave school for a week]! But still there was vibration!

This time it was coming from the other half of the building a thousand feet away where some contractors were working with a jackhammer repairing plumbing pipes in the concrete walls. They had to go too. Then there was the air-conditioning. We tried to shut off all of the vents in the lab and the surrounding rooms but to no avail. So, it was turned off. Next came our shoes. Everyone had to walk barefooted or in their socks. And then finally after four days of frustrating set up, we were ready.

Victor told us he had come up from Moscow just to observe the experiment; and with Konstantin and his staff of Professors, and ten other curious professors from different departments in the building, made the total of about 20 witnesses. A big commercial television camera, mounted on a huge three wheeled tripod, filmed the experiment. We could not be in the room because our breathing disrupted the results; so all the shoeless scientist witnesses [some without socks] were forced to stand about 50 feet away down the tile covered hallway and watch the experiment on a television monitor.

When I turned on the power and started the experiment, the screen that the split and then reassembled laser beams were focused on came into view. And when I slowly increased the electric current to the electro-magnets [with electronics set up by Prof., Nikolay Yur'evich Terekhin, EE; Leading Engineer of the Dept. of Atmospheric Physics], the Interference pattern began to change. While everyone sucked in their breath, I was able to quickly calculate that at maximum current to the magnets, the pattern revealed the speed of light had been *increased* by 204 mph! And then the argument started.

Everyone started yelling at each other at once. Because it was all in Russian I understood none of it. I just got out of the way, went down the hallway a bit to where Fabian and Victor were seated on a bench, sat down, and watched the spectacle. It seems, according to Victor's narrative, one group was saying I increased the speed of light while the other group said this is impossible: that I had only changed the frequency of the light. This discussion quickly turned into a violent argument that went on back and forth, unrelenting in intensity for about ten minutes until Konstantin finally had heard enough. The always composed and calm Konstantin suddenly lost his temper and yelled at them all to shut up and get out. Offended, almost all of them left, except for six of us.

One of the six was Dr. Lev Ivlev, Chairman of the Aerosol Physics Department [weather department] at St. Petersburg State University. He was one of the advocates arguing we had changed the speed of light and not the frequency. After the others left, he walked up to me with a big grin on his face and shook my hand so violently I thought he was going to shake it off. Then he motioned to all of us to follow him and took us down a couple of flights of stairs, and through a maze of hallways to his offices. He invited us into a smaller room that housed a large, extremely expensive table with ten matching chairs. The table had a thick glass top, beautifully inlayed with gilded art, showing scenes of ancient peoples and heroic battles from Russia's past; it looked like something right out of the Czar's dining room [perhaps it was?].

Next he took a huge key about six inches long out of a drawer, and used it to open a large double door safe about eight feet in height standing against the far wall. Inside it was surprising to see that this enormous safe contained only one object, a large bottle of American Whiskey! He then took expensive glasses also inlayed with gold from a drawer, and emptied the large bottle by filling them all up and passed them out to all of us. I was about to refuse because I don't drink when Fabian wisely warned me in a whisper, "Hey, we don't want to offend them, right?" Reluctantly Fabian and I, like everyone else, drank our full glasses all at once - with both of us coughing and choking - much to the amusement of the four professors. The professors were apparently→ "experienced drinkers (?)" who never once coughed or choked! Then after giving us chocolates to kill the bite of the whiskey, Dr. Lev stood up and made a speech in Russian while Victor interpreted:

"I have never seen an experiment like that in my life! What you did was not supposed to happen. Half of the people out there said what they saw was impossible. Nevertheless, it happened, so it is not impossible; I argued that only the possible can occur [where had I heard that expression before?]. But half the professors still did not want to believe what they saw. That's when they changed the argument, trying to say that you changed the frequency of the light and not its speed. But when I asked them to explain how the magnetic flux changed the frequency of the light, nobody could do it. Instead, they then started to argue that you are not a Doctor of Physics and have no right to be here in this establishment, and this experiment should never have been conducted!"

"So, I tell you what, next year I am holding my own international conference in Downtown St. Petersburg, and I am inviting you to come and present your revolutionary thesis as your dissertation for a PhD. You will have to be ready to defend it before the assembled and probably unfriendly scientists at the conference. I talked to Konstantin about you after he came back from America and showed me your résumé. And I am happy to say that with the two years of electronics schools you had in the military; your two year AA degree, and the large amount of advanced college credits equal to an additional three years of college you now possess from the California State University System, when it is all added up it amounts to seven years of school; you more than qualify for the First Degree in the Russian Academic system: the five year "Specialist Degree".

Next, because of all the additional work you have done explaining many of the most unexplainable mysteries of the universe that nobody else has ever done, such as Quantum Entanglement, and the Asymmetric Parity of Neutrinos; and furthermore, because of the two books you have written *The End of the Concept of Time*, and *The Vortex Theory* explaining many more mysteries nobody else has ever explained, you also qualify for the new master's degree program in the Russian Academic system: for the master's degree it takes at least two years of additional study, and a thesis of at least 70 pages in length. So, all the years you have spent on making your additional scientific discoveries [then about 80; now over 120], and the massive length and scope of the scientific fields of endeavor your discoveries encompass in fields such as Astronomy, physics, cosmology, and chemistry you more than qualify for a master's degree."

"And finally, the fact that Victor concurs and says you now have had more than five peer-reviewed scientific papers he has personally placed in international conferences throughout the world, [Note: in Russia you have to have at least 5 peer reviewed papers accepted by international science conferences in addition to your academic studies for a PhD: (in Russia, the Western academic PhD is called a "Candidate of Science and Engineering")], you now have met all of the qualifications for a PhD in nuclear physics. Congratulations!"

Chapter 20
I Defend the Thesis…

I was "blown away"! Three department chairmen from two of the most prestigious university's in Russia were willing to back me for a PhD! I didn't know what to say.

My eyes must have been a little glazed because Konstantin put his hand on my shoulder and said affectionately, "My good friend." After all these years of rejection and refusal by American scientists to even look at my work, this was a great day.

On the bus ride back to the hotel, I was floating on air. Mainly because of what Dr. Lev had said, [and of course the full glass of whiskey I drank]! And while Fabian went out "bar hopping" to the night clubs in St. Petersburg with a younger friend of Konstantin's, I sat in my hotel room and reflected back upon the events that had brought me to this present "State of Being" in my life…I thought about the incredible synchronicity of events that had started this quest so many years ago: the intense UDT training that turned me into an achiever; the encounter with the friend reading the book, *There is a River*; the stopping for a soda in the little store in Florida; the many diverse college and university courses in science I took; the discovery that time doesn't exist as a fundamental principle of the universe; the seven years it took to mathematically prove it; the insane conference in Fort Collins; meeting Victor, and Victor's subsequent unrelenting help; the arrival of Jeannie; the arrival of Konstantin and his marvelous transformation→ now believing in both the Vortex Theory and in Jesus Christ; and finally the arrival of another powerful advocate today, Dr. Lev Ivley. It was truly incredible. But no time to rest, there was now another important assignment to undertake: the dissertation! And I knew from the negative, unreasonable response of some of the arguing professors at the experiment, I had to be ready. And soon I was…

Dr. Lev's conference called *The Research Centre of Ecological Safety of Russian Academy of Sciences* given at Saint Petersburg, was delayed 6 months and took place on February 8-10 in 2007. So it was in February of 2007 that we went back to Russia, stayed at the same hotel, and enjoyed the pleasure of eating some more of that delicious Russian bread. There is no bread in the world like Russian bread; and when Russian cheese and Italian wine are added along with a few tins of sardines from the North Sea, it makes a delicious meal. Both Konstantin and Victor were there. Victor again came by train from Moscow to act as my "Second", and back me up and be my ally. Since Victor, Konstantin and Lev were all members of the Russian Academy of Sciences these were some powerful allies indeed!

My goal and work for the past year was to anticipate every question that might be asked at the dissertation, so I could give the appropriate response. Luckily, almost all of the questions were asked in Russian with Konstantin and Victor, interpreting. The placing into evidence the five peer review affirmations made by the professors in Moscow years before, and the additional affirmations made by Victor, Konstantin, and Lev were enough to convince almost all of the assembled professors that the proper approval of the thesis was already completed. And the subsequent boring task of the abbreviated repeating of a long question from Russian into English, and then the subsequent re-abbreviating process during the re-interpreting of my long answers from English back into Russian greatly slowed the process down, and [luckily for me] bored many members of the conference who wanted it all to end, and to go on to other business: all except for one disturbed professor!

Apparently, this professor from some far off province in Russia I had never heard of had just written a book on Relativity and wanted to hear nothing of what I had to say. He jumped up from his seat and called me every swear word in the Russian language until a big Russian Colonel, in full military uniform, [who had just given a speech before mine on global warming], got offended, stood up, grabbed the unruly professor by his neck and slammed him down into his chair. The professor was about to say something to the Colonel, looked up at his huge size, and thought better of it.

Luckily, because I do not speak Russian, I did not understand that he was cursing at me. So, I just shrugged my shoulders and smiled [this made him even madder]. Konstantin later told me that many of the professors there were impressed by my calm decorum. This further helped their acceptance of me being an equal; and helped with their final approval of the thesis.

After all of this outlandish and rude behavior exhibited by this supposedly "respectable" professor, Dr. Lev said we all needed to adjourn for break. We all [about 70 professors] got up and went into the adjoining cafeteria. Here, some of the professors who spoke English were eager to talk to me. However, in came the unruly professor and started yelling all over again and made another embarrassing scene for the other self-controlled Russian Academics gathered there. Victor, who is the consummate politician, grabbed his arm and led him away to a table in the corner of the room and calmed him down. Victor even ended up buying a copy of his book just to shut him up. [Victor told him some ridiculous story that I was going to have it translated into English!].Then something strange happened…

Someone tapped me on the shoulder. When I turned around, standing before me was a man impeccably dressed in a dark blue suit, [most of the professors at the conference wore Coats, ties and "jeans" which were the fashion of the day in Russia (jeans were very expensive in Russia)]. This man of obvious military bearing, physically fit with a military haircut, looking something like about 35 years of age, asked me in perfect English, "Mr. Moon, does your theory explain how to create anti-gravity technology?" Stunned, and before I could say anything, someone else asked me something and I briefly turned around; when I turned back, he was gone. Later, when I got back in the conference room, as I finished my speech, I looked for him and couldn't find him.

When I asked Konstantin about this incident, he became immediately concerned. He told me and Fabian to only stay in areas where there were other people and never walk alone on deserted streets! That there were secret efforts by the military to develop anti-gravity technology and they were willing to go to any extreme to get it! I was both impressed and frightened. This was the reason I never went back to Russia, even when offered a job by the Dean of the Physics Faculty to teach English Composition and Western Scientific Terminology to students training to be translators, who were going to be interpreting western scientific Journals from English into Russian.

Later, back in America, strange things began to happen. Both Fabian and I experienced strange noises on our telephones. I thought they were just technical bleeps. But Fabian, who was a computer expert, and had worked as a supervisor for Microsoft before he became a mortgage broker, said they were tapped. In the meantime, because nobody at the conference after an hour of debate could successfully challenge the mathematics of the thesis, its defense was deemed a success by Victor, Lev, and Konstantin. This allowed it to now enter the "*next*" phase in the Russian academic process before one could be awarded a PhD.

Unlike Western academics, where a doctorate degree is awarded by the university the student attends, in Russia, things are radically different. In Russia, after successfully defending one's thesis, and after having 5 papers published in international conferences throughout the world, the thesis is then sent out to twelve other universities and given to the department Chairman and one "Leading Engineer" of the branch of science deemed appropriate. Both of them in at least eight of the twelve universities have to approve it. This means at least 16 other professors have to additionally approve it. The process can take many months. Mine took over a year.

I was later told by Konstantin that many of the professors who had to evaluate the thesis did not like it because it directly conflicted with everything they had been taught to believe. However, they could not find any mistakes in the mathematics or in the use of the formulas I presented. So reluctantly, they had to approve it. But this was still not the end!

Finally, the thesis then had to go to the powerful Russian Ministry of Education in Moscow and a panel of between five to seven additional physicists had to approve it. Like all of the rest of the physicists it had been sent to, they didn't like it either, or the additional fact that I was an American! They told Konstantin, who was my advocate, they were going to deny it because I had failed to fulfill the one last requirement for a PhD in Russia: I didn't speak a foreign language! [Note: in Russia, all professors are required to speak at least one foreign language!] However, Konstantin was ready for them; he immediately told them, "Yes, he does speak a foreign language…he speaks English!"

Apparently, Konstantin turning their own requirements against them caught them off guard! Reluctantly there was nothing they could do, apparently that was how their legal statute was written ["…the candidate has to SPEAK a foreign language"]; so now, according to their law, they had to approve it: and then, and only then, afterwards change the law (otherwise it would be an ex-post-facto law illegal in Russia as well as America). Which apparently they did, because now, according to an amused Victor; you have to speak *both* Russian and a foreign language as a requirement for a Russian PhD!

And so, after a year of wrangling, the Russian Ministry of Education finally granted me a PhD in nuclear physics. Unlike the large diploma's given out to PhD's in America that become vanity wall decorations, I got a tiny one [4 inches by 14 inches, in a small leather carrying case. It was back dated to 2005, when the long process originally started when the proof of the thesis was made via the revolutionary experiment we conducted at the V. A. Fock Institute of Physics. Also, unlike America, in Russia, all PhD's are given an official state number: mine is KY 032771. And so, after many years, and just like the scarecrow in the Wizard of Oz: who became respectable after receiving a diploma from the wizard and was now somebody; I'm now somebody in the world of science!!!

It might shock some people to know that I never cared one iota about any academic degree whatsoever; not that I am ungrateful. But rather, all I ever wanted to know was the truth about the universe; and to let other people know that Jesus knew more about the construction of the universe than all the greatest scientists who have ever lived: including Mr. Albert Einstein!

PART VI
THE LEGACY OF THE VORTEX THEORY OF ATOMIC PARTICLES

Chapter 21:
The Legacy of the Vortex Theory of Atomic Particles...

The legacy of the Vortex Theory of Atomic Particles is simply stunning! No other theory in the history of science has had so many successes explaining so many mysterious phenomenon. The number of mysterious phenomenon of the universe the Vortex Theory has explained is without precedent; the number of great mysteries of physics, quantum mechanics, particle physics, astro-physics, chemistry, and cosmology the theory has explained has now passed 120! There has never been anything like this in the history of science!!! Some of them are listed below...

- *Unification of Newtonian Physics and Quantum Mechanics [Book 6]*
- *Discovery of the 5th force in nature: the Anti- Gravity Force!*
- *The Force of Gravity Explained*
- *The secret of Quantum Entanglement explained*
- *The Explanation of how single photons create Double Slit Interference Patterns*
- *The Explanation of The Pauli Exclusion Principle*
- *The Explanation of Dark Energy*
- *The Explanation of Dark Matter*
- *The explanation of "Tunneling"*
- *Dispelling the Myth of the Higgs Boson Particle*
- *The First ever Explanation of the Constant of Fine Structure: the 1/137 mystical dimension-*
   less number all of the greatest scientists of the past 100 years have tried to explain!!!
- *Explanation of the Asymmetric Parity of Neutrinos*
- *How the Particle and Wave Theory of Light is created*
- *The Explanation of the Striking Parallel Between Newton's Law of Gravity and Coulomb's Law*
- *An Explanation of the Creation of the Universe [unlike anything ever proposed before]*
- *An Explanation for how the Universe will end [unlike anything ever proposed before]*
- *The Explanation for the creation of the phenomenon of Time*
- *The Explanation for Time Dilation Effects at near Light Velocities*
- *The Explanation for length shrinkage at near Light Velocities*
- *Explaining the Mystery of Mass [no Higgs Boson is needed!]*
- *The Electromagnetic Force Explained [like never before]*
- *The Weak Force Explained [like never before]*

The Strong Force Explained [like never before]
How the Particle and Wave Theory of Matter is created
The Explanation of Intrinsic Spin [1/2 spin]
The Explanation of Newton's Three Laws of Motion
Resolving the Conflict between Inertial Mass and Gravitational Mass
The Explanation of the Conservation of Charge
The Explanation of the Conservation of Angular Momentum
Explaining the Conservation of Momentum
The Conservation of Mass and Energy is explained
The Explanation of the Mystery of Entropy
What the Neutrino really is!
The true Explanation of Buoyancy
The Explanation of how the Proton is created
The Explanation of how the electron is created
The explanation of how the Neutron is created
The Explanation of Covalent bond in Chemistry
The explanation of the Ionic bonds in Chemistry
The Mechanical Explanation of the Michelson Morley Experiment
The <u>Five</u> Forces in Nature are explained using configurations of space!
The Explanation of Mass
The Explanation of Energy
What causes Acceleration
The Explanation of the Muon's Prolonged Lifetime when moving at Relativistic Velocities
All Phenomenon Associated with the Theory of Relativity are now Explained
The explanation of the Strange Perihelion progression for the planet Mercury
Why all Particles possess the Same Amount of Charge
The Explanation of Black Holes
The Explanation of Planck's Constant
The Explanation of Increasing Velocity and Increasing Mass
The Grand Unification Theory
Why Electrons Orbit Protons
The True Vision of Space
How the Proton and the Electron create a hydrogen atom with two flowing vortices
The Vortices in higher dimensional space
Particle Collisions
The True Vision of Energy
The end of Einstein's Spacetime!
The End of the Theory of Relativity
Creation of the ±1 Charge,
The creation of Spin
Creation of the Up and Down Quarks
Creation of the Strange and Charm Quarks
Creation of the Top and bottom Quarks

"The Four Layers of Matter"
Leptons Explained
Neutrinos Explained
How Particle Collisions Create New Particles
The Explanation of how Quarks Change "Flavor"
The Explanation of the Law of the Conservation of Lepton Number
Lepton Creation during the Decay of Positive and Negative Pions
Neutrino Creation during the Decay of the positive Muon
Neutrino Creation during the Decay of the Negative Muon
The Decay of the Positive Muon
The Decay of the Negative Muon
The Collision between a Proton and an Electron Anti-neutrino
The collision between a proton and a Muon Anti-neutrino
The Collision between a Neutron and an Electron Neutrino
The Collision between the Neutron and the Muon Neutrino Collision
The Decay of the Neutron and the Creation of the Anti-neutrino
The Explanation of the Law of the Conservation of Baryons
The Explanation of the Conservation of "Strangeness"
Gauge Bosons are not Force Carriers between Particles
The Explanation of the CPT Theorem
The creation of the Pentaquark and the "Neutral Pentaquark"
The Motion of Photons and Particles through Electric and Magnetic Fields
The Reason why the Photon's Electric and Magnetic Fields exist; and why they are at Right
  Angles to each other
The Stability of Protons; the Instability of Mesons
The Explanation of what Quarks are
The Explanation of Quark Confinement
The Two Sides of Space!
The Two volumes of Space: one increasing; one decreasing
The REAL Explanation of the 1/3 & 2/3 Charges of Quarks
The Explanation of ±2 Charge of Resonances
The Explanation of 3/2 spin
The Explanation of the Up Quark
The Explanation of the Down Quark
The Explanation of the Strange Quark
The Explanation of the Charm Quark
The Explanation of the Bottom Quark
The Explanation of the Top Quark
The Explanation of the Muon
The Explanation of the Tau
The Explanation of the Electron, Muon, and Tau Neutrinos
The Difference between Strong Force and Weak Force Creations
The Explanation of how Quarks Decay into other Types of Quarks

The Explanation of "The Law of the Conservation of Lepton Number"
The Explanation of Neutrino Creation during Neutron Decay.
The Explanation of the "Law" of the Conservation of Strangeness
The Explanation of the Strange Quark's Extremely Long Lifetime
The Explanation of the W Particle
The Explanation of the Z Particle
The Reason why Electric and Magnetic Fields exist in Photons and why they are at Right
  Angles to Each Other!
The Explanation of the CPT Theorem
The Explanation of the Pentaquark.
The Proposal of a new Particle in Nature: THE NEUTRAL PENTAQUARK
Dispelling the Myth of Gluons
Dispelling the Myth of Gravitons
The Explanation of the Three Color Charges in Quantum Chromodynamics
The Explanation of Anti-matter
Creation of the Alpha particle
How the nucleus of an atom <u>creates</u> gamma rays and x-rays
How fusion creates energy
GOD! Has GOD been discovered!!!
Universal Religion?

Chapter 22
Legacy Continued: Quantum Entangled Technology
For Cell Phones Is Now Possible…

Until the discovery of the Vortex Theory of Atomic Particles the mechanism that creates quantum entanglement was unknown. Now that it is known the possibilities it possesses are almost unlimited. For cell phones, its abilities are nothing short of miraculous:

First, because the transmission travels through *fourth dimensional space*, it bypasses matter, allowing the transmission to "appear" to travel directly through the Earth! This ability reduces or eliminates most of today's infrastructure. Shockingly, it eliminates the need for the hundreds of billions of dollars worth of satellites, antennas, transmission lines, telephone poles, fiber-optic cables, and their associated amplifiers.

Next, it will create better reception. Because transmission takes place through fourth dimensional space, the limitations created by the obstructions of three dimensional matter are gone. People in mountainous terrains will not lose their signal. A sailor in a submarine in the Mariana Trench will be able to talk on a cell phone to a coal miner on the opposite side of the world in a West Virginia coal mine. When the NASA spacecraft goes behind the Moon, the signal will not drop out. Nor will it drop off for the Coast Guard or rescue workers in a Hurricane. No more people trapped in attics during a flood or storm will die because they cannot call for help on cell phones whose infrastructure is damaged. And finally, and most shocking of all, a person will be able to talk to an explorer on Mars or on one of the moons of Jupiter or Saturn in "Real Time": no time delays will occur, [see the *Pipe and Golf ball analogy* below*]!

Third, because the signal cannot be intercepted it is secure. This makes current secret low frequency RF transmissions to submarines in the Pacific Ocean obsolete. This helps save the whale's and other marine mammal's sensitive to sonar. Also, the transmission takes place instantaneously. This does not violate the speed of light's electromagnetic transmissions, but instead, tricks it. This effect is best described by having a long line of golf balls in a plastic PVC pipe…and leads us to the "*Pipe and Golf Ball Analogy* below"…

*The Pipe and Golf ball analogy:

Using the analogy of pipes and golf balls, if the speed of light is represented by a single golf ball put into a 20ft long empty pipe set on a 45 degree angle, and if it takes one second for the golf ball to roll all the way down and out of the pipe, this can be equated to the speed of light.

However, a pipe filled with golf balls represents quantum entanglement. And if the balls are packed so tightly that the mere act of putting a ball in one end causes another ball to instantly fall out of the other end, and if all one cares about is seeing a ball pop out of the end of the pipe, it appears as if the speed of light has been greatly increased: note: this experiment has actually been done using photons in a pipe. And just like the golf balls, the injection of a photon into one end causes another one to almost instantaneously "pop out" of the other end; making it appear as if the speed of light was increased by 300 times! [See

Scientific Paper titled → "*Detailed statement on faster-than-c light pulse propagation*": by Lijun Wang, Alexander Kuzmich, and Arthur Dogariu; NEC Research Institute; 4 Independence Way, Princeton, NJ 08540, USA.]

I told Fabian that this technology will eventually be needed by everyone: manufacturers of cell phones; manufacturers of transmitters and receivers; all fire, and police departments; rescue organizations; the four branches of the military and NASA. Eventually this superior method of transmission and reception will render all present transmission and reception devices obsolete. It will replace today's inferior transmission and reception technologies. It can also be used in place of batteries to charge cell phones. As a result, its value is incalculable. It can be sold outright for many millions, or licensed for billions! Fabian was ecstatic! As a Christian he wanted to generate enough money to feed the poor of the world!

That's all Fabian needed to hear; he was on a mission. He created a new company called CRYSTO Technologies [named for Christ and Technology], and he was on assignment to gain funding for this project. We added our four friends in Russia as consultants to the company out of compassion for their meager salaries as professors in Russia: [then, professors got free rent on a one bed-room apartment and around 6600 Rubles a month salary; the 33/1 exchange rate made their pay equal to 200 American dollars a month!] needless to say, they were behind us 100%.

So, while Fabian was creating websites and was out looking for funding, I drew up the schematics [thank you Navy ET School in San Francisco!] for a new type of cell phone transmission device we called the "4DQUID Secure Send & Receive Device". This device was to be incorporated into cell phones in place of their transmission and reception circuitry.

Next, a switching apparatus based upon Boolean Algebra [thank you Navy Crypto School!] was designed and the schematics drawn for this revolutionary electronic device that would allow the 4DQUID phone to switch modes and also place regular calls to the conventional cell phones of this era. Allowing it to do double duty: to act as a secure quantum entangled phone to call other quantum entangled phones; or act just as a regular phone like everyone else's. The work was done in secret, and I applied for no patents, because I did not want anyone to see the work [even someone at the patent office: who might be paid to give it to someone else?]!

Then came the hard part: finding and creating the quantum entangled chemicals needed to make the phones work.

The chemical I chose to use for the quantum entangled element was salt: sodium chloride. But I did not stop there. I also tried using different forms of salt: sodium bromide; sodium iodine; sodium fluoride; potassium chloride; potassium bromide, and several more. It took a year, but on December 31, 2007 at 1:47 PM we finally succeeded! We transferred photons of light from the sodium atoms to the chlorine atoms a distance of about 15 feet. It was a great day. But the effect only lasted for a little while. In searching for an answer, I finally found that the culprit was the humidity. Above 55% and the humidity was too great and interfered with the effect.

Months later we transferred the photons a distance of 11 miles from Fabian's mother's Real Estate offices to my house. This was another great day. However, I had not yet solved the problem of how to create the correct chemical balance to keep the phenomenon sustained for long periods of time when tragedy struck; Fabian got arrested!

All of the chemicals we purchased on the internet for our Quantum Entanglement Experiments were legal, but apparently the government did not think so. One day, Fabian received a knock on his door, when he answered, three policemen from the Broward County Sheriff's Department stood there [two men and one woman]. They said they needed to talk to him and asked if they could come in. Fabian, not knowing any better said yes.

When they walked in the trouble started. As two of them questioned Fabian about the chemicals, the third began to tear his apartment apart! He threw all of the bottles of soap and cleaning materials under the sink in the kitchen upon the floor. Then he went into Fabians bedroom and threw everything in the closet all over the room. Fabian, in a state of shock, did not know what to say.

Luckily or unluckily, Fabian's sister who lived two doors down saw the police car and showed up to see what was going on. When she saw what the police were doing to Fabian's apartment, she asked them if they had a warrant. The head policeman said they didn't need one because Fabian said they could search his place. When she asked Fabian if that was true, he said "NO"! and that's when the policeman said, "O.K. your under arrest!" When his sister said, "What for?" The policeman said, "He has a warrant out for his arrest because of a 10 year old fishing ticket he never paid!" Then he was handcuffed and hauled off to jail! [Later, we found out that his mother had paid for the ticket 10 years ago! That he was falsely arrested!]

When he was later taken before a judge, it was discovered that the Okeechobee Wildlife Ranger who wrote the ticket had retired years ago and was nowhere to be found! Hence, the charges were dismissed.

However, because Fabian was from Argentina and did not want to start trouble, he let it go and did nothing about it, even though he spent three days in jail!

What is really a joke if one can be made out of this outrageous and illegal act by the police is that all of the boxes housing the chemicals we had purchased were sitting right outside the front door to his apartment! If the police were not careful, they would have tripped over them! [Note: Fabian's apartment was really an addition to his father's house; a converted car port renovated by his father years before; and his front door opened onto the covered driveway under which all the chemicals were stored!]

Next, the police came to my house!

That same day the three policemen came to my house, knocked on the front door, and asked for me. When Jeannie said I was in the back yard, they came around the side of my house, entered the screen door to the patio and confronted me! When I asked them what they were doing, the lead police officer gave me his card [I still have it!] – *Detective Steve Robson,* of the *Strategic Investigations Division* of the Broward County's Sheriff's Office– and arrogantly told me, "We are here to determine if you are a terrorist or starting a methamphetamine lab!"

Then without even asking, this policeman then sat down at my table and proceeded to open up a folder that had a computer printout of everything Fabian and I had purchased on the internet. When I asked him if anything we had bought was illegal, he said "No!" So, I said, "Just a minute" and called Fabian to get his list of chemicals. When his phone rang and rang, none of them said anything. Nor did they tell me he had just been arrested.

In hindsight, because *they never even told me my rights*, I should have told them to leave, or have told them that if they wanted to question me they had to do it in front of a lawyer. But they said nothing about why they had been secretly watching me. However, never having dealt with the police before, I was curious as to why they were here. It was also very curious to note that they knew I had been to Russia, had printouts of my trips, and asked if I spoke Russian!

Because I had nothing to tell them they appeared disappointed and left. But the lesson was learned: our government cannot be trusted: and the police will violate your rights if given the opportunity! I suddenly became distrustful of a government I had always respected!!!

Chapter 23
Legacy Continued: Anti-Gravity Engineering Is Now Possible…!!!

In the winter of 2011, using the principles of the Vortex Theory of Atomic Particles, one of the greatest scientific discoveries ever made took place during a *secret* experiment in South Florida: the discovery of the fifth force in nature – *The Anti-gravity Force*. [The word *secret* is emphasized because of the outrageous experience of the police illegally investigating my perfectly legal quantum entangled chemical experiments made me go "underground."]

Here, I must say that ever since I had discovered the Vortex Theory of Atomic Particles, I suspected this fifth force in nature existed around electrons, but I did not know how to prove it. However, one unusual phenomenon I had witnessed aboard ship in the Navy was lightning strikes going directly into large waves of the Atlantic Ocean during stormy conditions when I was at sea. This phenomenon had always intrigued me because somehow the salt water had to have been able to absorb the massive amount of electrons in the lightning bolt.

Many years later, during research I was conducting charging salt water with static electricity, I suddenly realized how an experiment could be conducted that would allow an anti-gravity field surrounding the electron to reveal itself. Instantly, I dropped what I was doing and focused all my attention solely on designing and building an apparatus to test this theory. And, after several months of trials, it was successful beyond my wildest dreams!

It worked! What a thrill! Suddenly, I and I alone was privy to one of the great secrets of the construction of the universe: the existence of a fifth force in nature, an anti-gravity force! I told Fabian about the work, and he took all my experimental data from a hundred tests and used it to create the necessary graphics for the paper I subsequently wrote.

I was leery of telling anyone about it because I realized the shocking implications of such a discovery, and wanted to keep it as secret as I could for as long as I could.

Out of courtesy, I told Konstantin about the discovery and sent him a copy of the paper. Much to my horror, he was so excited about it he showed it to his closest colleagues Professors Lev, and Nikolay. They were so enthralled by it they published it in the January 2012 branch of the *Journal of the Russian Academy of Sciences*, sponsored by St. Petersburg State University in Petergof.

I didn't ask them to do this, nor did I even consider they would do something like that because I wanted to keep the discovery quiet. When I kind of casually [so as not to offend him] asked Konstantin about why they published it, he said, "…because we are your friends."

Luckily, many Russians speak English, but they do not read English; just as many Americans speak Russian but do not read it. Consequently, many Russian journals are left unread by the West; so luckily, not many people on either side of the Atlantic have read the article. Which is very important, because after realizing its implications, I wanted to be the scientist who developed this technology and have control over whom I gave it to! And "I must say… Wow!… What a fantastic technology it is!" There has never been anything like it. It will revolutionize life upon this planet. Nothing will ever be the same again…

Just as the discovery of the internal combustion engine created the worldwide revolution that put the horse and buggy out of business, so anti-gravity technology will put the internal combustion engine and all of the technology associated with it out of business too. Incredibly, because this anti-gravity technology is all based upon the tiny electron, these anti-gravity engines are all based upon electronics!

In the past, the effects of the electron's anti-gravity force have gone unnoticed because of the *massive* disparity between the slight strength of this force and the huge strength of the electron's electrostatic force. And although this discovery will create a revolution in the world of physics, creating havoc with the Standard Model; it will create an unbelievable revolution in the current technology's that dominate the way of life for the people of the Earth. It will create the world's next industrial revolution!!!

When my friends in Russia asked me what this anti-gravity technology was capable of doing I told them the following…it begins with fossil fuels becoming obsolete…

In the future, electronic circuitry alone powered by batteries or capacitors will be used to propel a new generation of aircraft and spacecraft! For example: anti-gravity fighter "aircraft" will be able to fly faster and out maneuver anything any other country currently has. _Two different types of modified anti-gravity technology_ allow both the creation of smaller "fighter" aircraft, and huge, gigantic transport aircraft capable of carrying thousands of men!

The discovery of the anti-gravity force will finally allow the creation of fantastic new technologies that will revolutionize all military aircraft. For example: the best of today's military aircraft are no match for a "craft" that can fly at speeds of thousands of miles per hour, and then, while moving at this same speed, avoid missiles or bullets by making instant sharp right angle turns to avoid being hit and without the pilots feeling any G forces whatsoever!. Also, if pursued by large numbers of aircraft, it can merely fly upward into space and rain missiles down upon the opposition from this higher strategic location.

These military implications are simply incredible! No sonic booms will reveal the presence of the craft! Amazingly, it can stay on station for weeks at a time without "refueling;" only needing to come back to base to rearm itself and replenish supplies for the pilots. And of course, the most shocking of all: <u>all</u> of this era's military aircraft, from jet fighters to huge transport planes become obsolete!

One can easily take off from Australia, go into space, carefully circumnavigate half the world in 30 minutes, and land at a base located in Germany. Since it can also take off directly into space, then traverse the distance from one country to another while flying hundreds of miles above the earth and then descend directly out of space towards a target on the ground, no aircraft carrier is needed, nor are bases needed in foreign countries. Such a disparage between it and today's craft is equivalent to having an F16 fighter going back in time, and fighting the biplanes of the First World War. There is no competition.

Commercial aircraft using anti-gravity technology will also be superior to anything currently flying. Commercial "aircraft" will be able to fly anywhere in the world for a nominal cost. The two types of technology also allow the creation of the small, two seat type of craft for individuals; intermediate forty to three hundred passenger craft; and large five hundred to a thousand passenger craft.

Its logistics is also beyond belief: for example, noiseless aircraft will silently lift off directly upward into the air from downtown parking lots and fly anywhere in the world at supersonic speeds. Other craft will easily fly to the Moon or Mars at 1/10 light velocities; and the colonization of our solar system suddenly becomes possible!

Space sickness, physical deterioration, and bone loss are also eliminated. A second set of anti-gravity devices can be placed within the superstructure of the craft to create artificial gravity! Hence, voyages to Mars will become sightseeing trips for the retired or elderly!

The dual ability to travel long distances at tremendous speeds, and then land upon a small pad placed upon a downtown skyscraper makes the helicopter obsolete. Today's large airports with long runways will no longer be needed: they can be turned into golf courses!

Fuel costs will dramatically decrease because the propulsion system will not burn fossil fuel. The environment will improve, and when this technology is eventually applied to the automobile the shocker is this: there is no more automobile! Just as rubber wheels made the wooden wagon wheel obsolete, and put blacksmiths and livery stables out of business [and there were literally tens of thousands of them two centuries ago], this technology will put rubber wheels out of business. The only ones that will probably be kept will be for future "nostalgic" races such as the 24 Hours of Le Mans; or the Indianapolis 500. [Note: just as we still have horse races today, we will probably still have car races in the future.]

Using anti-gravity technology for the generation of electrical power, global warming will come to an end. The carbon footprint is gone! Jet fuel is gone! Just think, no jet fuel; no runways; aircraft can land anywhere; and upon landing, no refueling trucks and ground crews needed; weather is unimportant, wind no longer a factor. BUT MOST IMPORTANT OF ALL, this technology can be used to turn generators of any size such as those at Boulder Dam!

However, and shockingly so, today's dams and electrical power plants will eventually become obsolete and go out of business too! One day, just as electricity replaced the candle and kerosene lamp technology of the 19th century, antigravity powered generators in every home in the world will free mankind from the reliance and domination of the giant utility companies that presently exist! Because they do not give off carbon dioxide gases, small anti-gravity generators will provide the electricity for every house in the future! Today's power lines will no longer be needed!

Again, space implications are simply incredible. Large rockets needing huge amounts of fuel are suddenly obsolete. The cost per kilogram of supplies sent into space becomes trivial. Food and supplies for the space station can be loaded into an anti-gravity craft sitting in a Wal-Mart parking lot, and then flown directly into space! NASA would suddenly be downsized. Huge rocket launching pads are no longer needed. Cape Canaveral is obsolete. Russian rockets are no longer needed to send astronauts into space from the Baikonur Cosmodrome. Just like driving cars, astronauts can "commute" into space in their own crafts and park at the station on a daily basis, if they so choose!

The later version of this technology will also make the present day vehicles and cranes used for lifting and moving objects obsolete: the same goes for energy production. To say this is revolutionary is an understatement: Cranes and derricks are no longer needed. Houses can be built in central locations, and then flown to far off locations. Devices placed under ships will allow them to fly *over* the Panama Canal! Private individuals will be able to pilot their own spacecraft to the Moon or Mars!

Consequently, after telling my Russian friends *some* of the above, they finally agreed that my extraordinary PhD thesis had finally found extraordinary proof. Originally, they did not want to publish my thesis because I was first asked to find physical proof of this theory. I was told that it has to be extraordinary proof: that an extraordinary thesis such as mine, requires extraordinary physical proof! And when my friends in Russia saw the proof, after eight years of waiting, they finally agreed to publish my thesis and added the anti-gravity paper with it.

Using analogies, if gravity is described as a concave depression in space, anti-gravity would then be described as a reverse convex bend in space. This effect can be created using electronics. Many countries in both the West and the East, and private companies such as Boeing, have been trying to develop anti-gravity technology for years. All have failed because they do not understand the construction of space as outlined in the Vortex Theory of Atomic Particles.

It must also be mentioned that anti-gravity technology supersedes and overcomes the disadvantages encountered by today's new "Hypersonic" technology. The hypersonic technology presently being developed is limited by the heat generated friction of the air upon the surface of craft. New materials are needed to endure the 4000+ degree Fahrenheit temperatures. These high temperatures result from the fact that present forms of aircraft "split the air" apart in front of it; forcing air molecules apart during the passage of the craft. This creates friction of the air upon the materials making up the craft. Anti-gravity technology does not work like this!

Instead of "splitting" the air in front of the craft, anti-gravity technology "splits the space" in front of the craft. This reduces friction because the craft is not touching the air atoms and molecules that are imbedded in the space. Hence, the generation of heat is avoided. However, because the configuration of the atoms in front of the craft do not form a uniform structure as in say a crystal, consequently, some collisions with atoms are unavoidable. However, the heat generation is miniscule in comparison to that encountered by today's hypersonic technology. Anti-gravity technology is so revolutionary that there is nothing to compare it to at this time.

It must also be noted that anti-gravity technology usurps and negates the hypersonic technology today's military's of the world are trying to develop. These hypersonic rockets that are purported to travel at five times the speed of sound and cannot be intercepted, are in effect going slow in comparison to the speed that anti-gravity propelled craft can travel at.

Even more important, an anti-gravity craft would be able to intercept and deflect an incoming asteroid that could destroy a part of the earth's surface or completely destroy life upon the earth as the massive asteroid did that wiped out the dinosaurs sixty five million years ago. It could reach the incoming asteroid in almost no time at all, and gently nudge it off of its course just enough to change its trajectory and miss the surface of the earth.

It is easy to see that anti-gravity technology will usher in a new era of possibilities that were unobtainable or unthinkable just a few years ago.

Like the start of a gold rush, although many companies, universities, and individuals will attempt to be the first to develop this technology, at present, they lack the fundamental knowledge as to why this revolutionary force exists; they also lack the understanding and the explanation of its most singular and bizarre characteristic: (the secret to developing anti-gravity engineering) – something that can only be called, "The Magnificent Dilemma"! For example, when the electron is within the atmosphere of the earth, why does it seek to escape from the gravity of the earth? Why does its anti-gravity effects cause it to move upward into the upper atmosphere where it creates the spectacular lighting effects known today as Sprites, Jets, and Elves? And yet, when it is out in space, why does the gravity of the earth attract it? The answer to this vitally important question is absolutely crucial in creating anti-gravity technology…"

Although a clever individual carefully studying all of the Vortex Theory of Atomic Particles discoveries might be able to discover the answer to the above question, it has been left out of all six of these books for good reason! The person that develops this technology must not be one whose intention is to enslave the world, but rather, to make it a better place for all mankind.

An example of such a person is Stan Clifford.

Chapter 24
I Meet Stan Clifford…

Fabian's efforts to find investors for his company CRYSTO Technologies brought forth what seemed to be a never-ending stream of con-artists and crooks. I ended up having to deal with them and having to tell Fabian that they were either misrepresenting themselves, or were out and out crooks looking to first learn, then later steal the technology.

Because I wanted nothing more to do with these people, I determined to develop the technology myself even if it left me financially in debt. Then Fabian told me about meeting Stan Clifford. Initially I did not want to talk to him because I thought he was just another con-man. But then Fabian told me how he met him. It was most fascinating…another case of synchronicity.

It seems that Fabian was taking some courses at Miami Dade Junior College in order to become an Ultrasound Medical technician. In one of the classes he was in, he met a student named Jamie. Because both were older than the other students, they developed a mutual comradery and ended up having a few minor conversations together before class. However, one afternoon, when Fabian went out into the parking lot, his car would not start. Luckily, he spied Jamie getting into his car, yelled at him about needing a battery jump, and this started it off.

Grateful for his help, Fabian told him about his ultimate plans to create Vortex Theory Technologies to charge cell phones; and after listening, Jamie told him about an acquaintance of his named Stan Clifford who might be interested in such discoveries. Stan owned a paint company in Stanford Kentucky that made specially colored acrylics for artists all over the world. He had a yacht in Miami and enjoyed coming down to South Florida for vacations in the winter.

So, to make a long story short, Fabian eventually met Stan; told Stan about me, and eventually I met Stan. And when I did, I was impressed. Stan was a man of my generation, brilliant of mind, yet whose talents have been applied in a different direction than mine: mine in science and acquiring knowledge; his in business and making money.

Stan can only be described as one of those special people who have done everything right in life. He went to college right after high school, played sports, and graduated with a degree in business from Clemson University, in South Carolina; then later got an MBA from Minnesota State University at Mankato. He then used this business knowledge to start a company he made very successful, and eventually became a member of Clemson's entrepreneurial senior advisory group. However, he was also deeply interested in science, something he had missed out on, having concentrated on business curriculum when at college.

When we talked, I told him directly what I was doing so if he was not interested it would end further dialogue right there and not waste the time of either one of us. I told him about how I acquired my PhD based upon the words of Jesus in the New Testament. How I wanted to reveal to the world that Jesus knew more about the construction of the universe than all of the greatest scientists who have ever lived including Albert Einstein. That I wanted to make the world a better place for all people everywhere and what better way to do it than to develop anti-gravity technology. That such technology would free mankind from fossil fuels and allow us to leave the planet and travel to the stars. Furthermore, that I believed in UFO's because of the 1980 incident at Rendlesham Air Force Base in England witnessed by Lt. Colonel Halt and two of his soldiers: John Burroughs and Jim Penniston who both had actually stood beside the craft for over one half hour when it was sitting upon the ground in the midst of Rendlesham forest! [Note: Penniston drew in his notebook, both the shape of the craft and the geometric symbols on the outside of it that appeared to be some sort of language.]

I told him that I also believed in the 1965 Kecksburg incident in Pennsylvania where the entire town's fire department observed the strange "acorn shaped craft" that had crashed upon a farm outside of town. In fact, they were there standing beside it until the Air Force arrived, kicked them off the property and took over the site. Afterwards, it was loaded upon a flatbed truck, covered with a big canvas then driven through town and out of the area in the middle of the night.

Amazingly, this did not turn Stan off either or make him think I was crazy [as some professors at universities did who are afraid to talk openly about UFO's]!

In fact, Stan was all for everything I said, especially the part about Jesus. He had been raised as a Catholic, where time, and living in this century of technological miracle innovations had not diminished his beliefs in God as had many of our generation. He was willing to fund the development of anti-gravity technology. Because Fabian and I had different agendas: his to develop quantum-entangled batteries to charge cell phones; and mine to develop anti-gravity technology, we ended up going our separate ways [for a while that is].

Then a strange sequence of events took place…It is almost as if evil was trying to keep me from succeeding.

I had only been working for a few months upon anti-gravity technology, when I discovered my little lab behind my house was bugged. I discussed this during a face to face meeting with Stan, and we decided to not say anything. If it was the government, and they discovered that we knew they had bugged the lab, they might invoke National Security and try to stop the research. However, if it had been one of the private groups that had contacted me, [one such group was made up of retired college professors and former CIA and NSA agents], then we could use this as an opportunity to hold false conversations in the lab and lead them astray.

Another strange event occurred one evening while practicing meditation. Many years ago, while attending college at California State University in Long Beach, I learned how to meditate when I was studying Hinduism and visited the Ramakrishna Monastery in the hills above El Toro Marine Corps Air base in Southern California. I practiced this form of meditation in my lab in the evening when it was quiet and there were no interruptions. The purpose of this meditation is to quiet the physical mind, then like a traveler going to some far off land, think of something you would like to know and then await to see what enters your mind. I practiced this meditation to open my mind to the scientific technology I was seeking to develop. But during one episode I encountered something totally unexpected.

In all of the many years I have practiced meditation I have never encountered the presence of another entity within my mind! Yet that is exactly what happened!

Although many know me as the cold blooded scientist who deals only with scientifically proven facts and writes only technical papers, somehow, a beautiful tale of erotic love and horrific war invaded my mind, grabbed hold of my conscious thoughts, and would not leave me alone.

I cannot explain how this happened except to theorize, that perhaps sometimes when the conscious mind of an individual is standing upon the frontier of science at the edge of the known universe of thought, the door to the subconscious mind is opened wide and what might enter might not be what was invited. This appears to have been the case.

Apparently, the souls of two powerful individuals – Gorgo and Leonidas – the Queen and King of ancient Sparta, entered into my mind and shared their modern reincarnated story with me. I suppose I should be honored because Leonidas is famous as the leader of the 300 Spartans, who fought the million man Persian army at the battle of Thermopylae 2500 years ago. Nevertheless, my thoughts and thought processes were seriously interrupted by this remarkable individual and his charming wife.

In all honesty I do not know if they exist somehow as disembodied souls, reincarnated spirits, or if they were somehow just products of my imagination. However, none of that matters because they were alive in my mind, possessed my consciousness, and would not relinquish control of my thoughts until I agreed to tell their story. Because I could not continue to work until they left, as bizarre as it might seem, I made a curious bargain with them. Although some might try to argue that I made a contract in my mind with my mind, just the same, the deal was struck, and I told them that they would not pass into oblivion. Instead, they could depart from my mind and live forever in the pages of world literature. They agreed, the bargain was struck, and I finally regained control of my thoughts!

Amazingly, this short book, under 100 pages, was written in only 11 days, and with almost no editing! The story was transmitted into my consciousness, almost as if I was watching a movie in my mind. I just wrote as it came to me, and at the end I entitled it, *"The Girl with the Mona Lisa Smile."* When it was over, the two consciousnesses left my mind never to be heard from again.

So shocked was I by this incident, I researched other authors and scientists and found that a number of them have had similar experiences: most notably, Robert E. Howard, a prolific writer, most famously known for his *Conan the Barbarian* series of short stories. Stories he apparently experienced much like mine. He was once quoted as saying, "That he just had to write them down whenever they came to him, wherever he was, and before they left his mind and were gone forever."

Robert Louis Stevenson also spoke of talking to his "Brownies" in his mind: little hooded beings in long brown robes who talked to him and told him stories. His famous story *The Strange Case of Dr. Jekyll and Mr. Hyde* came through him in a dream he did not instantly forget like most other dreams because of being prematurely awoken by his wife. He spent three days locked in a room writing notes for the future famous novel, and to keep this story from being forgotten.

Perhaps the most famous scientist who has had a similar experience was Thomas Edison. I actually visited his laboratory [now a museum] in Fort Meyers Florida where I saw the bed he would lie down upon to take quick naps so he could communicate in his mind with "his" Brownies who gave him ideas for his inventions.

In 1713 the composer Giuseppe Tartini had a dream where he talked to, and then heard the Devil playing a sonata on his [Tartini's] violin. He immediately awoke and tried to replay as much of it as he could. He later wrote down the music and called it the *Devil's Trill Sonata*. [It is also known as the *Violin Sonata in G Minor*]. Giuseppe later lamented to his friend the renowned French astronomer Jérôme Lalande that there was a great gap between the beautiful sonata he heard and what he was able to remember and write down later.

Some of the most famous of these visitations by other entities are those documented by the famous 19th Century poet William Blake. William Blake's dead brother Robert who died of tuberculosis in 1887 when William was 31 years old, would come to him in his mind and give him poems and stories to write. After completing a seven year apprenticeship, William eventually became a journeyman copy engraver, and worked on projects printing engravings and books. A year later, Robert is credited with coming into William's mind and giving him a new method of printing his works that William called "illuminated printing." This method allowed Blake to control every aspect of the production of his art.

Others who credited dream experiences and talking to entities in their minds for their advice were: Beethoven, Richard Wagner, and Edgar Allen Poe, just to name a few.

After learning all of this knowledge, suddenly, I did not feel all alone, or that I had gone crazy. Nevertheless, after this incident, I wanted to get out of South Florida and decided to accept Stan's invitation to go to Stanford Kentucky and set up my lab in one of his factory's buildings. This went well for a while until another strange incident occurred: my laptop was stolen out of my lab and then later returned!

When I set up a lab, I use the old Navy axiom, "Everything has a place and everything in its place." As such, every piece of electronic gear in my lab has a certain place reserved specifically for it. It starts there in its place at the beginning of the day and is put back in its place at the end of the day. And so it was with the laptop.

After being in Stanford for a couple of months, during an idle conversation with Stan on the phone, I informed him I was going to put all of my data on the laptop I had recently purchased and thought nothing more of it. Then about a week later, I decided I needed to wash some clothes and went to the coin laundry about one mile away. I was gone for only about an hour and a half, but when I returned I went to get the laptop and it was not there! I searched high and low for it. I even looked in my car; then drove back to Stan's house in Danville where I had been staying and determined it was not there either. Back at the lab I took a coke out of the cooler and sat back in my chair to think about where it could be when I inadvertently glanced to my left and saw it sitting upon a metal shelf under a pile of papers and cables. Then the alarm bells went off in my head.

I knew I did not put it there. The cables belonged upon another set of shelves and the papers belonged in my files! So somebody else had to have done this: but why? Why would they take it then bring it back? There could only be one reason, they wanted to copy the data on it. Someone had to be monitoring either me, or Stan's phone to realize this! When I told Stan about this, he came over to the lab. Neither one of us could figure out how someone could get into the lab because Stan and I had the only two keys. However, Stan's brother David who is a brilliant engineer and runs the factory also came over to the lab and almost instantly noticed how the pins in the hinges on the recently painted door were not seated properly; they were sticking up about ½ inch above the hinges and had no paint on them! This indicated that they had been pried up and taken off, allowing the door to be swung open from the hinge side without unlocking the door handle on the other side. The rug in front of the door also revealed marks in it caused by two paint spatulas found in a nearby closet that were used to swing the door out over the rug. Flecks of paint were also seen on the rug!

To say I was outraged is putting it mildly. But the more important question was, "Who could have done this?" There were other electronics gear in the lab that were worth thousands of dollars. But none of this was stolen. Only the laptop worth 300 dollars was taken. Hence, it was mutually concluded that only someone who wanted to copy the contents of the laptop would have taken it then returned it; and hid it in the lab to make it look like it had simply been misplaced.

There is one ironic twist to this story, the laptop had the new version of Microsoft Word in it, and I could not figure out how it worked. So the laptop had no data in it whatsoever!!!

In trying to determine who could have done such a thing I realized that there were many possibilities: our government was number one. However, one other intriguing possibility was the Chinese! They were working on anti-gravity research, and several months before, I had been invited to China to speak at an international conference on quantum entanglement. The email I received said I had been chosen because of two articles I had written for a certain on-line scientific journal. However, I had never written any articles for this journal!

I also remembered hearing about an American Physicist, a woman of Chinese descent, Dr. Ning Li who was working on anti-gravity research, had gone to China for some reason and was never heard from again! Some say she is now back in the United States and is working secretly for the DOD, but nobody knows for sure. What is known is that her research has disappeared and her colleague, Dr. Larry Smalley, the former Chairman of the Physics Department at the University of Alabama at Huntsville, has died and his research has disappeared as well! Suddenly, anti-gravity research did not seem to be very fun anymore.

Other companies and individuals working on anti-gravity technology have gone quiet too…

- ❖ Dr Podkletnov, a Russian physicist who originally discovered the loss in weight above super-conducting coils went underground back in Moscow for years while working for an undisclosed corporation, at an undisclosed location with only a phone attached to an answering machine taking messages for him.
- ❖ Boeing Corporation was also secretly working on anti-gravity.
- ❖ The military wing of the UK hi-tech group BAE Systems is working on a quiet anti-gravity program, dubbed Project Greenglow.

- NASA is also secretly attempting to reproduce Dr Podkletnov's findings.
- The military's of China; the United States; Germany; France; India; and Japan are also working secretly on it too.

Everyone suddenly appeared to be going quiet on anti-gravity research! The question is why? Who or what are they afraid of?

This was enough for me. Knowing the secrecy everyone working on anti-gravity research was keeping, I realized it must be a good idea. I realized I needed to be in a place where I knew who belonged in the area. That if a stranger was present, they would look out of place, and be identifiable. So, I packed up everything and went back to Florida.

Here, I had many successes. I built an "apparatus" under which I was able to create a 3-5% weight loss! However, it was unsustainable. Without going into details, the elements I used kept shorting out, nullifying the results. Thirty-five times I tried different configurations of shape and electronics. I kept making modifications but still the electronics kept shorting out.

I eventually rented a warehouse in West Palm Beach because I needed to build a special oven made of bricks on a concrete floor to be able to heat tungsten up to 1200 degrees Fahrenheit. But this did not work because the plastic the apparatus and electronics were made of was too close to the tungsten and even though protected by asbestos, it would nonetheless melt and destroy the experiment.

[Note too: the lab was protected by special infra-red cameras installed to catch any intruders. Also, the main part of the warehouse was kept in disarray to lead anyone who casually looked in to think that nothing out of the ordinary was taking place. Also, false materials and false "apparatus" were left in the lab to confuse any intruders.]

I did have success with the theoretical aspects of the Vortex Theory of Atomic Particles. I discovered that there has to be a fourth charged Lepton, I called the WOW particle because of its great mass: see Book 3: *The Explanation of the Quark Theory 2021*. It is too bad both Konstantin and Victor are not still alive [both had died of heart attacks!]. Both would have been ecstatic to hear of such a discovery.

Another discovery they would have been happy to see was the explanation of the Constant of Fine Structure. This magic dimensionless number of 1/137 had been a mystery for 100 years. Although most people outside of science have never heard of it, it is nonetheless of great importance. Without this number, carbon nuclei would never be able to form in stars and there would be no life on earth or anywhere else in the universe that uses carbon atoms as the foundation for life.

I consider the discovery of the explanation for the Constant of Fine Structure to be one of the crowning achievements of the Vortex Theory of Atomic Particles. As explained in Book 2, "Since its discovery by Arnold Sommerfeld in 1916, the Constant of Fine Structure has been one of the great unsolved mysteries of science." Perhaps it would not be that important to the world's scientists if it had been discovered by anyone other than Dr. Sommerfeld. Dr. Sommerfeld was an extra-ordinary scientist and a "Super Educator". And when I say Super, I mean SUPER!!!

He trained many of the greatest scientists of his day; seven of which went on to win the Nobel Prize: Werner Heisenberg, Wolfgang Pauli, Peter Debye, Hans Bethe, Linus Pauling, Isidor I. Rabi, and Max von Laue.

Therefore, it seems curious to today's scientific community that neither Sommerfeld nor any of his famous students could solve the mystery of a dimensionless number whose value is 1/137!

It is also shocking to learn that not only Sommerfeld and his group of famous Nobel Prize winning students, but many of the other most famous 20th Century scientists have tried and failed to solve the mystery of the Constant of Fine Structure: Albert Einstein, Werner Heisenberg, Arthur Eddington, Carl Jung, Niels Bohr, Richard Feynman, Arthur Eddington, I. J. Good, Enrico Fermi, J. Robert Oppenheimer, Edward Teller, and Stephen Hawking just to name a few. However, perhaps it was Nobel Prize winner Richard Feynman who explained it best when he stated…*"It has been a mystery ever since it was discovered more than fifty years ago, [actually 100 years now] and all good theoretical physicists put this number up on their wall and worry about it. Immediately you would like to know where this number comes from: is it related to pi or perhaps to the base of natural logarithms? Nobody knows.* **It's one of the greatest damn mysteries of physics: a magic number that comes to us with no understanding by man."**

Nevertheless, the efforts to find a solution to this conundrum have not ceased. Numerous "Quests" to find a mathematical solution to this strange dimensionless numerological constant continue today. However, the failure of the greatest 20th Century scientists who have tried and failed to solve this problem for the past 100 years is not their fault. The answer* to the mystery of what the Constant of Fine Structure represents has eluded all who have attempted to solve it because they possessed an incorrect vision of how the matter, space, time, energy, and the forces of nature of the universe are constructed.

[*Note: the mystery of the Constant of Fine Structure is solved when it is found to be the Sine of a very small angle: 0.48^0! It is a dimensionless number because the Sine, Cosine, and Tangent of all angles are dimensionless. See Book 2: *The Vortex Theory of Atomic Particles: Part II*]

Chapter 25

Flashback to the Past…

I Discover the Secret of How to Interpret Jesus' Words, Statements, and Parables From His Own Point of View, as if He Were Standing Right in Front of Us Explaining Them to Us!

Every once in a while, the pool business in South Florida would "dry up"! Interest rates would go high, making it hard for people to get loans for luxury items such as swimming pools, and the construction business inevitably suffered. Such was the case in 1982. Pool work being slow, I eventually got a job in electronics repairing transceivers for a little company in Fort Lauderdale called Sunair electronics. Talking to other technicians working there, I heard about overseas electronics jobs available for licensed electronics technicians.

I researched the companies offering these jobs, and contacted some of them. Most of them didn't need anybody, however, one such company - called Kentron International - wanted to hire me immediately because of my military experience repairing receivers and transmitters and the secret Crypto Clearance I once had. It seems they had several Electronic Tech jobs available in the Marshall Islands for former military veterans. The only catch was they had to pass the federal test for an Electronics Communications Repair License: the government license that allows you to lawfully repair and align the powerful transmitters used by all radio stations in the United States. I said, "No problem."

For one intensive month, I reviewed all my communications electronics; went to the government offices in Los Angeles California and took the test for a "General Radio-Telephone Operator License," and passed [License # PG-11-2653]. Got hired, and was immediately flown to Hawaii by commercial airliner. Then from Hickam Air Force base in Pearl Harbor, I boarded a C-9 MAC Flight headed for the Island of Kwajalein in the South Seas; the infamous site of a horrific battle that took place between the Marines and the Japanese in WWII.

This island is at the other end of the Vandenberg Missile Test Range, and is the largest of the 97 narrow little islands that form a thin necklace of land completely encircling the bright blue waters of the Kwajalein Atoll: the largest atoll in the world at 70 miles by 50 miles; and one of the shallowest, whose huge central lagoon has a maximum depth of only 130 ft; making it easy for divers to retrieve the warheads shot as payloads upon the missiles fired out of Vandenberg Air force base in California.

All of the civilian contractors working there were former military veterans - like myself - with security clearances. It was just like being back in the navy but without officers and chiefs telling you what to do. [You were already expected to know how to do what you were assigned to do without needing supervision.]

My job was to repair and calibrate various communications gear on the island, including all of the transmitters, receivers, and transceivers located upon the Tugboats and the shallow draft LST's based in the harbor. During my free time, since there was nothing to do, I exercised, played baseball, worked out with weights and most important of all studied the strange, puzzling statements made by Jesus in the New Testament. Perhaps "studied" is an inadequate choice of words, maybe being perplexed better describes my attitude… I say this because many years ago, these were the quotes made by Jesus in the New Testament that not only confused me but directly conflicted with the teaching of every church group I ever talked to while at college. Perhaps these quotes should be called the shocking statements of Jesus…!!!

Shocking Statements of Jesus!

Although most Christians are happy to be able to quote most of Jesus' more popular statements, there are other statements no one ever talks about. These troubling statements are shunned, ignored by priests and their constituency. Different churches have conflicting explanations about them. [Some have even gone so far as to say that these cannot be Jesus' words at all, but rather, are misinterpretations created during the original translations from one language to another.]

But before I list them, and the questions they raised, it is important that I first mention that even though I was confused then by these quotes, looking back in retrospect, I realize how fortunate I was. If I had been trained in a Christian religious doctrine or discipline during the process of growing up, this study would never have taken place. In all likelihood, I would have accepted then, as truth, the erroneous explanations for these statements which would have been given to me by the various authority figures representing this particular religious denomination.

Because of youthful trust and naivety, I would naturally have assumed that these priests, teachers, and parents would have known what they were talking about. As a result, my curiosity would have been satisfied and I would never have felt the need and hence the motivation to seek any other explanation than the one with which I would have grown up with. Consequently, when I was older, I would unknowingly have a strong emotional attachment to my early training and would have found it extremely difficult to reject these explanations.

However, this was not the case.

Because I had no knowledge and no explanations regarding these statements, and since I was as of yet unaware of the previous insight, I soon found myself going from church to church and from priest to priest asking questions and seeking explanations for these peculiar statements and parables which I did not understand.

Furthermore, since I was not raised in a church environment, I naturally did not have any prejudicial inclination towards any one church organization or another. Therefore, I innocently and indiscriminately visited most of the various Christian denominations: Catholic, Protestant, Methodist, Baptist, Non-denominational, etc. Seven of Jesus' most notably and strange quotes stand out above the others: these are the statements and the questions I asked about them…

Question #1

If according to Jesus, love is the way one should feel for all people including one's enemies, then why did he make this unusual statement regarding those we should love the most?

Luke 14:26

"If any man come to me, and hate not his father, and mother, and wife, and children, and brethren, and sisters, yea, and his own life also, he cannot be my disciple…"

Question #2

If Jesus is called the prince of peace by those who teach Christian doctrine, how come his own words contradict this same doctrine?

Luke 13:51

"Suppose ye that I am come to give peace on earth? I tell you nay; but rather division: For from henceforth there shall be five in one house divided, three against two, two against three…"

Question #3

Why did Jesus in the following statements speak out not only against the ways of life which are natural, normal, and necessary for the continual propagation of the human species, but also, that which is necessary for its very survival?

A. Against sexual relations [for men]:

Matt 19:11

"All men cannot receive this saying, save they to whom it is given. For there are some eunuchs, which were so born from their mother's womb: and there are some eunuchs, which were made eunuchs of men: and there be eunuchs who have made themselves eunuchs for the kingdom of heaven's sake. He that is able to receive it, let him receive it."

B. Against the natural and normal instinct to conceive and bear children [for women]:

Luke 23:28

"Daughters of Jerusalem weep not for me, but weep for yourselves and your children. For behold the days are coming, in the which they shall say, blessed are the barren, and the wombs that never bare, and the paps which never gave suck."

C. Against marriage itself: which is both the natural and normal instinct for males and females to pair off with one another to reproduce and continue the species.

Luke 20:34

"The children of the world marry and are given in marriage. But they which shall be accounted worthy to obtain that world and the resurrection from the dead, neither marry nor are they given in marriage: neither can they die anymore: for they are equal unto the angels; and are the children of God, being the children of the resurrection.

[Also, what do the words mean "…neither can they die anymore:" can a man die more than once?]

D. And against the most important of all instincts – the survival instinct itself!

Luke 9:23

"If any man will come to me, let him deny himself, and take up his cross daily and follow me. For whosoever will save his life shall lose it: but whosoever will lose his life for my sake shall save it."

Question #4

And finally, one of the most strange and mystifying of all his statements, is the one where Jesus states that John the Baptist *is* Elias! It is important to note that he did not state that John is like Elias [as some groups try to say] or that he is similar to him in some way, but rather, in Matt 11:9, Jesus states that John is Elias. [Equally important, he does not make this statement to all men but only to those willing to listen!]

Matt 11:9 (Jesus speaking about John the Baptist)

"But what went ye out for to see? A prophet? Yea, I say unto you, and more than a prophet. For this is he of whom it is written, behold, I send my messenger before thy face, which shall prepare the way before thee. Verily I say unto you, among them that are born of women there hath not risen a greater than John the Baptist: not withstanding he that is least in the Kingdom of Heaven is greater than he. And from the days of John the Baptist until now the Kingdom of Heaven suffereth violence and the violent take it by force. For all the prophets and the law prophesized until now. And if ye will receive it, this *is* Elias which was for to come. He that hath ears to hear, let him hear.

As I review these quotes, I recall how my questions were ill received when I boldly asked for explanations about them to the various ministers and priests of the different Christian denominations I visited while attending college. Apparently, these were quotes most of these religious teachers avoided talking about. Some even got offended, giving me a firsthand view of the dilemma of Christian doctrine in the process. I found that every time I asked a question, depending upon whom I asked, I got a different answer! To both my amazement and disappointment, not only did I get different and conflicting answers for all of the meanings of each of these individual statements: some even told me that these were not even Jesus' words [Imagine that (!), priests saying that their own Bible is wrong!]!

I also got different and conflicting answers to the meaning of Jesus' message as a whole! And some of these Priests, Bishops, Evangelists, Pastors, & Ministers, actually got quite upset being asked embarrassing questions they had no answers for. Eventually I gave up on asking others for explanations and began looking for some way to determine Jesus' own interpretation of his words? Was there one? (Yes there is!) But it took many long years of searching to find it!

One other question that intrigued me the most of all, and not on the list, was why Jesus had to go out into the wilderness to be tempted of the Devil? Was he not the Son of God? Was he not the most perfect man who ever lived? If so, then what was the purpose for these temptations he had to undergo? Was there some fault within him he had to overcome?

What did he have to prove? This bothered me. Amongst a host of other questions I wanted answered, this one became first and foremost. In its answer I felt certain it most likely possessed some of the other answers I was looking for.

It also created an unfortunate incident when casually talking about Jesus' quotes to some of the other civilian contractors working at Kwajalein. One of them, an atheist, was offended by anything religious and asked this hostile question [and in a hostile manner]: "How do you even know that Jesus was this so-called Messiah? How do you know it was not Mohammed, or some other religious leader like Buddha?"

Taken by surprise, I found his question hard to believe. I never heard it asked before, especially in such a rude and discourteous manner; and did not know how to answer it. It was only after I compiled the following list of quotes from the Old Testament prophesying the coming of the Messiah that I could checkmate this atheist's questions, and the questions of all others who did not believe Jesus was the Messiah…

> In looking for an answer to the question of who the Messiah was, it is understood that not only is the Old Testament of the Bible a collection of books containing the history, laws, mythology, and religious philosophy of the Jewish people, it also predicts the coming of a Messiah. Because both the other major religions of the world [Christianity & Islam] also use the Bible to justify their beliefs in *their* Messiah, we investigate it to see what it says. And as we search through the Old Testament for quotes referring to the Messiah we find many.
>
> Some refer to where he will be born; others refer to the mysterious return of the famous Old Testament prophet Elijah who will precede him as a herald proclaiming him to be the Messiah. Others mention what the Messiah will do, what he will say, what others will do and say to him; and then there are those that refer to his unspeakable death.
>
> These prophecy's referring to the Messiah's birth, life, and death can be found all through the Old Testament Books of the Bible. Remarkably, they coincide perfectly with the life of Jesus as told in the New Testament of the Bible! In fact, they coincide so exactly, they can be considered to be a perfect synopsis of his life! However, the above choice of descriptive words is probably too mild: these paranormal statements from the distant past are so shockingly accurate about Jesus' life they are more adequately described as being, uncanny, supernatural, and mystical!!!
>
> Whose life in the history of the world has ever been foretold or prophesized like this…
>
> <u>Where he will come from</u>
>
> *Deuteronomy 18:15-16 [<u>Moses</u> speaks of Jesus] "The lord thy God will arise a prophet from among his fellow Israelites";*
>
> *Jeremiah 31:31 The lord will make a new covenant with the people of Israel;*
>
> *Samuel 7:12-13 The Messiah will be the offspring of David;*
>
> *Isaiah 7:14 A virgin will give birth to him;*
>
> *Micah 1:2 He will come from the clans located in Bethlehem;*
>
> *Hosea 11:1 He will be called back from Egypt;*

Elijah will return (?), proceed him, and herald his coming

Isaiah 40:3-4 The prophecy of the return of Elias [as John the Baptist] heralding the coming of Christ;

[Note: the name Elijah and Elias are used interchangeably in Isaiah, Malachi, and Matthew.]

Malachi 3:1 "Behold I will send my messenger [Elias as John the Baptist], and he shall prepare the way before me;"

Malachi 4:5 "Behold I will send you Elijah [as John the Baptist] the prophet before the coming of the great and dreadful day of the lord;"

(Note too: When John the Baptist was in prison, he sends two of his disciples to Jesus to ask if he is the one they are looking for and waiting for? And in Matthew 11:4-6 Jesus replies to them; and then in Matthew 11:7-13 Jesus speaks to the masses about the greatness of John and finishes in Matthew11:14, by stating that John is in fact Elias (!): *"And if ye will receive it, this is Elias that was forth to come. He that hath ears to hear, let him hear."* Note too: the last part of his statement reveals that this saying is not meant for all people, but only for those capable of understanding it!)

What he will say and do

Psalm 78 1-2 He will speak in parables about the secrets of the earth;

Isaiah 6:9-10 He will speak to people who will not understand;

Isaiah 8:14 He will become the "stone" that causes people to stumble;

Isaiah 9:1-2 He will be a great light to those living in darkness;

Isaiah 11:10 His lineage from Jesse [David's father] will rally the nations to him;

Isaiah 35:5-6 "The blind will see, the deaf will hear, and the lame will be healed";

Isaiah 42:1-4 He will bring justice, but he will be non-violent;

Isaiah 53:3 He will be despised and rejected;

Isaiah 61:1 The lord has anointed him to free the captives;

Isaiah 22:22 He is the key to the house of David;

Daniel 7:13-14 All nations of the world will worship him;

Zechariah 9:9 He will enter Jerusalem riding on the foal of a donkey;

He will be betrayed

Psalm 31:13 They conspire against him and plot to take his life;

Psalm 38:12-13 [Again as above], they plot his ruin;

Psalm 41:9 His close friend [Judas] will turn against him;

Psalm 31:11 Everyone he knows will reject him;

Zechariah 11:12-13 He will be betrayed for 30 pieces of silver;

He will be beaten and crucified

Isaiah 50:3-6 [Jesus being scourged] He offers his back to those who beat him, his cheeks to those who pulled out his beard, and his face to those who spit on him;

Psalm 22:8 He is mocked [on the cross by the people]: "let him [God] deliver him;"

Psalm 22:1 Jesus cries out [on the cross]: "My God, my God, why hast thou forsaken me?"

Psalm 22:13-15 Jesus thirsts [on the cross];

Psalm 22:16 [Jesus nailed to the cross] "They pierced my hands and feet;"

Psalm 22:18 [Jesus on the cross] "They parted my garments among them, and cast lots upon my vesture"

Psalm 69:21 [Jesus on the cross] they gave him vinegar for his thirst;

Psalm 31:5 [Jesus says on the cross] "Into your hands I commit my spirit;"

But he will not die but live!!!

Psalm 118:17-18 "I will not die but live…the Lord has chastened me severely, but he has not given me over to death;

[And many more, though not as explicit as these.]

Shockingly, for Mohammed, we find nothing!

How can this be? According to Islamic doctrine, Mohammed was a great prophet and God's messenger sent to present and confirm the teachings preached previously by Adam, Abraham, Moses, Jesus, and other prophets. In fact, many of the prophecy's concerning the Messiah were written by Isaiah in the book of Isaiah; and by David in the book of Psalms. So how come none of these prophecy's by these two authors refer to Mohammed, his life, his death, and his ascension to heaven? How come we hear nothing about these holy places of Islam such as Kaaba and Al-Masjid al-Haram, in Mecca; or Al-Masjid an-Nabawi in Medina?

Instead, the quotes we do find refer to the Messiah's place of birth, life, and death in places like Bethlehem, Egypt, Judea, and Jerusalem. Nothing about Mecca or Medina or the Holy places listed in the Koran! Furthermore, none of the famous locations that refer to Mohammed's birth, life, and death are even mentioned in the Old Testament except for Jerusalem where Jesus was crucified. The prophecy's describing the horrific death of the Messiah refer directly to Jesus' trial and crucifixion, and not about Mohammed's simple death when he fell ill and died. Also, the quotes about what the Messiah will say and do refer to Jesus' subsequent quotes and acts, and not to those of Mohammed. Nor was there a herald proceeding Mohammed, proclaiming him to be the Messiah, as the prophesies of *Isaiah* and *Malachi* foretell.

> Consequently, using the principles of the scientific method, because the overwhelming evidence concerning the prophecies of the Messiah in the books of the Old Testament refer directly to Jesus' birth, life, and death; including his statements and his acts as described in the New Testament, we find that the parallel is nothing short of supernatural, uncanny, and unmistakable – leading us to conclude that Jesus had to be the Messiah the Old Testament of the Bible was prophesying of.
>
> It should also be noted that neither the births, life stories, nor deaths of Buddha, Lau Tzu, Zoroaster, Confucius, or Rama or any other religious leader, compare to any of the prophecy's describing the birth, life, and death of the Messiah in the Old Testament.

Needless to say, those who scoff at Jesus being the Messiah are brought to task by these Old Testament prophesies made hundreds of years before his birth and have nothing to say! I have also heard it argued that Jesus knew of these prophesies and exploited them by speaking in parables, and insisting upon riding into Jerusalem on the foal of a donkey. But if this was true, how about those prophesies he couldn't fake? How did Moses know he would come from among his fellow Israelites? Or how did Samuel know he would be a descendent of David? How did Micah know he would come from the clans in Bethlehem? How did Hosea know that the Messiah would be called out of Egypt? Or how did Isaiah know he would be of the lineage of David's father Jesse? Or hundreds of years before it happened, how did Zechariah know he would be betrayed for 30 pieces of silver? In other words, if he was a fraud, how could he have faked all these things he had no control over?

Nobody has been able to answer any of these redirected and retorted questions!

But even more important, when I realized that Jesus' description of the location of the Kingdom of God could only be located in fourth dimensional space; and using this hypothesis, the scientific study and its subsequent mathematical analysis resulted in the discovery of the existence of a pure fourth dimensional space with *no time characteristics*! In light of this astounding scientific discovery, and the even more astonishing way it was discovered via only one quote by Jesus in the New Testament, we are forced to come to the even more astounding conclusion: that *Only the Messiah, speaking for God, would know exactly where the Kingdom of God is located*! Nobody else on earth, either then or now, knew of such a location, not even Mr. Albert Einstein!

THIS GREAT DISCOVERY MARKS THE END OF THE OLD DYSFUNCTIONAL ERA WHERE SCIENCE AND RELIGION WERE CONSIDERED TO BE SEPARATE ENTITIES, AND THE BEGINNING OF A NEW ENLIGHTENED ONE: A NEW ERA WHERE THE BELIEF IN GOD RETURNS AND OVERPOWERS THE DOMINANCE OF THE BELIEFS OF THOSE WHO BELIEVE IN NOTHING! A NEW ERA WHERE HOPE AGAIN JOYFULLY ENTERS THE HUMAN HEART, WHERE SCIENCE AND RELIGION BLEND TOGETHER AND BECOME ONE; THE WAY IT WAS ALWAYS MEANT TO BE!

Chapter 26
Working 24/7 for a Year, I Finally Make the Analysis!

Luckily, after 18 months, I was soon back in the states, happy to get off of that claustrophobic, cramped little island, and enjoy some wide open spaces. In fact, the idea of the wide open spaces sounded good, really good; so, I built a wooden "over the cab camper" on my old pickup truck and took off! I drove all over the western states from California to Washington; to Montana; then down to New Mexico, Arizona, and finally ended up in Northern California. However, all the while traveling and enjoying seeing the West, I was nevertheless continually thinking about the seven questions and Jesus' mysterious need to go out into the wilderness. I knew there *had* to be some way, some method via which I could determine exactly what Jesus meant by each and every one of these statements. Like Woody Guthrie, or Vachel Lindsay, who worked on their music or poetry while wandering through America, my wanderings were not idle traveling either, but were filled with purpose.

I eventually lost the camper! It was only nailed onto the wooden bed, and got blown off in a big dust storm in the middle of Monument Valley Utah. At first I was temporarily angry at a friend of mine who recommended it be nailed on to the wooden flat bed of the truck by 16 penny nails. He told me it was all right to nail it down because he had just passed his state contractors exam in California and knew exactly how much force a 16 penny nail was capable of withstanding [he never considered the effects of vibration]. However, later, after I thought about this unfortunate incident, I could only thank him. Because if the camper had been thoroughly attached with bolts as I had originally planned, the whole truck would have been blown off of the road, just as a large 18 wheeler had done up ahead of me! [Was this also synchronicity at work?]

A family of Navajo Indians from the reservation located there, seemed to come out of nowhere [they had probably done this before!] and helped me load my possessions and I gave them the camper that was still in pretty good shape considering the accident it was in. Their kids enjoyed picking up the two thousand odd pennies I told them they could have but were now scattered all over the highway when the large pickle jar they were in was smashed.

I then ended up staying in cheap motels. After a year, the $10,000 dollars I had saved had almost run out. But that was all right. During this year of sightseeing, and visiting national parks while contemplating the mysterious statements of Jesus, I sadly realized I was still no closer to the truth. I also realized that even though I went all over the West, I had gone nowhere! I realized it had been a fruitless, futile search: a waste of time! Reluctantly, I stopped at another little motel in Redding California and reviewed what I had done so far…

…I had studied all of the religions of the world. I had gone to University, studied Physics, Geology, Chemistry, Astronomy, Mathematics, Anthropology, Archeology, and all of the scientific theories regarding the creation of the universe. I spent time in monasteries – of several different faiths. I chanted, meditated, and prayed. I did everything I could think of. But at the end, I was no closer to the truth than when I began.

There was only one option left…do what Jesus' said in Matthew 19:21…"If you would be perfect, go, sell what you possess and give to the poor, and you will have treasure in heaven; and come, follow me."

So, in the spring of 1985, when I heard from the local Ranger Station that enough of the snow had melted in the mountain passes for backpacking; I gave away all the money & possessions I had in life, was baptized at the little Inter-Denomination Church in Orville, and then, with nothing but the clothes on my back, went up into the Trinity Mountains of Northern California to meet God or die.

This was not suicide. This was not a case of running away from something, but rather, of running towards something. I also knew from studying near death experiences that sometimes at the point of dying, some people have had the profound experience of seeing Jesus and knowing the answer to every question, about everything they ever wanted to know. And after so many years of searching, if it only happened at the point of death, then so be it.

However, it was not meant to be. It only took one night of lying on my back and staring at the stars turning in the heavens above to have an experience: a profound and painful understanding, that I was not yet fit to meet God.

I was not fit because I had no love in my heart. I had no good deeds to offer in exchange for my soul. I lacked compassion.

It was a bitter experience.

But without knowing it then, I had somehow done exactly what was needed to accomplish the goal I had now sought for so many years. For when I came back down out of the mountains, a rush of new thoughts began to fill my mind. I began to experience what can only be called, *"The Knowing, the Understanding."* I call it this because there is no other way to describe it.

Amazingly, all I had to do was think about a question I had regarding Jesus' words and soon the answer would appear in my consciousness. It can only be described as coming from within you and through you. I remember saying time and time again, "Ah yes, yes"; "It is so simple, so simple". But even more important: *The method –"the philosophical key"– of how to make a study to determine exactly what Jesus meant by his Shocking Statements also came into my mind. I now knew exactly how to do it: but when could I do it???* Unfortunately, I had more pressing matters to attend to.

I was broke, and needed to make some money; and just like the answers coming into me, almost magically, this problem was suddenly, and "synchronistically" solved!

During a phone call to a friend, I was told about a construction job in Southern California that had just opened up. The man I had given my old truck to, gave it back, plus some money to get down to Southern California. I got the job and within six months, I had saved ten thousand dollars again. By working in the day and guarding the construction site at night, I was able to quickly make enough money to quit, and go to Kingman Arizona. Here

I lived for a year in an inexpensive motel. Working 24/7, I was finally able to make the analysis of the words of Jesus in the New Testament I had waited for, for so many years.

Using the philosophical key - I had discovered in the mountains - I was able to apply it to learn the explanation for <u>all</u> of Jesus' words, statements, and parables (including his "Shocking Statements") as if he were standing in front of us explaining them to us. In the history of mankind, there has never been a "Scientific/Religious" study like this ever. This ground-breaking vision of the universe, revealing man's place and purpose for being here, is simply stunning, revolutionary!

Perhaps the word "revolutionary" is an inadequate word. This knowledge is so revolutionary there is nothing in the world to compare it to! It stands alone and apart, unique from all the other knowledge ever discovered about mankind. It reveals our purpose in life, and the one of two diametrically opposed, completely different destinies that await us all…

…the destiny of mankind is split between two different groups of souls: those who will save themselves and those who will destroy themselves. This analysis presents the mental thoughts, emotional feelings, and actions necessary for those who will save themselves via "the way of the soul": shockingly there will not be many, "Matt 7:14…narrow is the way, which leadeth unto life, and few there be that *find* it!"

Then are the mental thoughts, emotional feelings, and actions of those who will destroy themselves via "the way of the animal"! This will be the majority, "Matt 7:13…for wide is the gate, and broad is the way, that leadeth to destruction, and many there be which go and only a few find it."

As such, for those who will save themselves, it is a tale best remembered, a wonder to behold; but for those who will destroy themselves, it is a tale of terror…[or hopefully (?), maybe a tale needed for the necessary repentance, and reversal of destinies?] Either way, it is…

…………………."The Secrets of Life and Death" ……………………..!!!

The end

References

National/International Conferences attended, and peer reviewed scientific papers presented

[1]   The Vortex Theory of Matter. [Presentation of his own work]
      'International Forum on New Science' Colorado State University (1992, Sept 17-20).
      Moon. R. Fort Collins, Colorado. USA. Topic: The Vortex Theory of Matter. Copyright 1990)

[2]   The Vortex Theory and some interactions in Nuclear Physics. [Book of abstracts; p. 259]
      'The LIV International Meeting on Nuclear Spectroscopy and Nuclear Structure; Nucleus 2004'
      (2004, June 22-25). Moon, R., Vasiliev, V. Belgorod, Russia.
      http://nuclpc1.phys.spbu.ru/nucl/Abstracts/Nucleus_2004.pdf

[3]   The Possible Existence of a new particle: The Neutral Pentaquark. [Book of materials; pp. 98-104]
      'Scientific Seminar of Ecology and Space' (2005, February 22). Scientific Research Centre for
      Ecological Safety of the Russian Academy of Sciences. Moon, R. Saint Petersburg, Russia.
      https://spcras.ru/ensrcesras/

[4]   Explanation of Conservation of Lepton Number. [Book of materials; p. 105]
      'Scientific Seminar of Ecology and Space' (2005, February 22). Scientific Research Centre for
      Ecological Safety of the Russian Academy of Sciences: Moon, R., Vasiliev, V. Saint Petersburg,
      Russia.
      https://spcras.ru/en/srcesras/

[5]   Explanation of Conservation of Lepton Number. [Book of abstracts; p. 347]
      'LV National Conference on Nuclear Physics' (2005, June 28-July 1). FRONTIERS IN THE
      PHYSICS OF NUCLEUS. Moon, R., Vasiliev, V. Russian Academy of Sciences. St. Petersburg
      State University. Saint Petersburg, Russia.
      http://nuclpc1.phys.spbu.ru/nucl/Abstracts/Frontiers_2005.pdf

[6]   The Possible Existence of a New Particle: the Tunneling Pion. [Book of abstracts; p. 348]
      'LV National Conference on Nuclear Physics' (2005, June 28-July 1). FRONTIERS IN THE
      PHYSICS OF NUCLEUS. Moon, R., Vasiliev, V. Russian Academy of Sciences. St. Petersburg
      State University. Saint Petersburg, Russia.
      http://nuclpc1.phys.spbu.ru/nucl/Abstracts/Frontiers_2005.pdf

[7]   The Possible Existence of a New Particle in Nature: the Neutral Pentaquark.
      [Book of abstracts; p. 349] 'LV National Conference on Nuclear Physics' (2005, June 28-July 1).
      FRONTIERS IN THE PHYSICS OF NUCLEUS. Vasiliev, V. Moon, R. Russian Academy of
      Sciences. St. Petersburg State University. Saint Petersburg, Russia.
      http://nuclpc1.phys.spbu.ru/nucl/Abstracts/Frontiers_2005.pdf

[8]   The Experiment that discovered the Photon Acceleration Effect. [Book of abstracts; p. 77]
      'International Symposium on Origin of Matter and the Evolution of Galaxies' (2005, Nov 8-11).
      Gridnev, K., Moon, R., Vasiliev, V. New Horizon of Nuclear Astrophysics and Cosmology.
      University of Tokyo, Japan.
      https://meetings.aps.org/Meeting/SES05/Content/273
      https://flux.aps.org/meetings/bapsfiles/ses05_program.pdf

[9]   The Conservation of Lepton Number. [Paper presentation]
      'American Physical Society 72nd Annual Meeting of the Southeastern Section of the APS' (2005,
      Nov 10-12). Moon, R., Calvo, F., Vasiliev, V. Gainesville, FL. USA. APS Session BC
      Theoretical Physics I, BC 0008
      https://meetings.aps.org/Meeting/SES05/Content/273
      https://flux.aps.org/meetings/bapsfiles/ses05_program.pdf

[10] The Vortex Theory and the Photon Acceleration Effect. [Paper presentation]
'American Physical Society; March Meeting; Topics in Quantum Foundations' (2006, March 13-17). Gridnev, K., Moon, R., Vasiliev, V. Baltimore, Maryland. USA.
Abstract ID: BAPS.2006.Mar.B40.6
https://meetings.aps.org/Meeting/MAR06/Session/B40.6
http://meetings.aps.org/link/BAPS.2006.MAR.B40.6

[11] The St Petersburg State University experiment that discovered the Photon Acceleration Effect.
'American Physical Society; March Meeting' GENERAL POSTER SESSION (2006, March 13-17). Gridnev, K., Moon, R., Vasiliev, V. Baltimore, Maryland. USA.
Abstract ID: BAPS.2006.MAR.Q1.146
https://meetings.aps.org/Meeting/MAR06/Session/Q1.146
http://meetings.aps.org/link/BAPS.2006.MAR.Q1.146

[12] The Neutral Pentaquark.
'American Physical Society; March Meeting' GENERAL POSTER SESSION (2006, March 13-17). Moon, R., Calvo, F., Vasiliev, V. Baltimore, Maryland. USA.
Abstract ID: BAPS.2006.MAR.Q1.147
https://meetings.aps.org/Meeting/MAR06/Session/Q1.147
http://meetings.aps.org/link/BAPS.2006.MAR.Q1.147

[13] The Neutral Pentaquark. [Paper presentation]
'International Workshop on "Nuclear Physics with RIBF' (2006, March 13-17).
Vasiliev, V., Calvo, F., Moon, R. RIKEN Research Institution. Saitama, JAPAN.
Abstract: RIBF-Pentaquark.
https://ribf.riken.jp/RIBF2006/

[14] Nuclear Structure and the Vortex Theory. [Paper presentation]
'International Workshop on "Nuclear Physics with RIBF' (2006, March 13-17).
Moon, R., Vasiliev, V. R. RIKEN Research Institution. Saitama, JAPAN.
Abstract RIBF-Vortex
https://ribf.riken.jp/RIBF2006/

[15] Experiment that Discovered the Photon Acceleration Effect. [Paper presentation]
'International Workshop on "Nuclear Physics with RIBF' (2006, March 13-17).
Moon, R., Vasiliev, V. R. RIKEN Research Institution. Saitama, JAPAN.
Abstract Moon 1
https://ribf.riken.jp/RIBF2006/

[16] To the Photon Acceleration Effect. [Paper presentation]
'APS/AAPT/SPS Joint Spring Meeting' (2006, March 21-23).
Moon, R. San Angelo, Texas. USA. Abstract ID: BAPS.2006.TSS.POS.8
https://meetings.aps.org/Meeting/TSS06/Session/POS.8
http://meetings.aps.org/link/BAPS.2006.TSS.POS.8

[17] The Saint Petersburg State University Experiment that discovered the Photon Acceleration Effect. [Paper presentation] 'American Physical Society; Astroparticle Physics II' (2006, April 22-25). Gridnev, K., Moon, R., Vasiliev, V. Dallas, TX. USA. Abstract ID: BAPS.2006.APR.J7.6
https://meetings.aps.org/Meeting/APR06/Session/J7.6
http://meetings.aps.org/link/BAPS.2006.APR.J7.6

[18] The Photon Acceleration Effect. [Paper presentation]
'American Physical Society; Session W9 DNP: Nuclear Theory II' (2006, April 22-25).
Gridnev, K., Moon, R., Vasiliev, V. Dallas, TX. USA. Abstract ID: BAPS.2006.APR.W9.6
https://meetings.aps.org/Meeting/APR06/Session/W9.6
http://meetings.aps.org/link/BAPS.2006.APR.W9.6

[19] The Neutral Pentaquark. [Paper presentation]
'American Physical Society; Session W9 DNP: Nuclear Theory II' (2006, April 22-25).
Moon, R., Calvo, F., Vasiliev, V. Dallas, Texas. USA. Abstract ID: BAPS.2006.APR.W9.9
https://meetings.aps.org/Meeting/APR06/Session/W9.9
http://meetings.aps.org/link/BAPS.2006.APR.W9.9

[20] Controversy surrounding the Experiment conducted to prove the Vortex Theory. [Paper presentation] 'American Physical Society; 8th Annual Meeting of the Northwest Section' (2006, May 18-20). Vasiliev, V., Moon, R. University of Puget Sound. Tacoma, Washington. USA. Abstract ID: BAPS.2006.NWS.C1.9
https://meetings.aps.org/Meeting/NWS06/Content/518
https://meetings.aps.org/Meeting/NWS06/Session/C1.9

[21] The Photon Acceleration Effect. [Paper presentation]
'American Physical Society; 8th Annual Meeting of the Northwest Section' (2006, May 18-20). Moon, R., Vasiliev, V. University of Puget Sound. Tacoma, Washington. USA.
Abstract ID: BAPS.2006.NWS.C1.8
https://meetings.aps.org/Meeting/NWS06/Content/518
https://meetings.aps.org/Meeting/NWS06/Session/C1.8
http://meetings.aps.org/link/BAPS.2006.NWS.C1.8

[22] Experiment that Discovered the Photon Acceleration Effect. [Paper presentation]
'International Congress on Advances in Nuclear Power Plants' ICAPP '06, (2006, June 4-8).
Gridnev, K., Moon, R. Reno, Nevada. USA. American Nuclear Society.
Abstract 6006. ISBN: 978-0-89448-698-2

[23] The Neutral Pentaquark. [Paper presentation]
'International Congress on Advances in Nuclear Power Plants' ICAPP '06 (2006, June 4-8).
Vasiliev, V., Calvo, F., Moon, R. Reno, Nevada. USA. American Nuclear Society.
Abstract 6045. ISBN: 978-0-89448-698-2

[24] Is Hideki Yukawa's explanation of the strong force correct?
'The International Symposium on Exotic Nuclei' Book of abstracts: Joint Institute for Nuclear Research. (2006, July 17-22). Vasiliev, V., Moon, R. Khanty Mansiysk, Siberia. Russia.
http://wwwinfo.jinr.ru/exon2006/
http://jinr.ru/

[25] The Explanation of the Pauli Exclusion Principle. [Paper presentation]
'59th Annual meeting of the American Physical Society Division of Fluid Dynamics' (2006, Nov 19-21). Moon, R., Vasiliev, V. Tampa, Florida. USA.American Physical Society;
Abstract ID: BAPS.2006.DFD.P1.17
https://meetings.aps.org/Meeting/DFD06/Content/578
https://meetings.aps.org/Meeting/DFD06/Session/P1.17
http://meetings.aps.org/link/BAPS.2006.DFD.P1.17

[26] Is Hideki Yukawa's explanation of the strong force correct? [Paper presentation]
'59th Annual meeting of the American Physical Society Division of Fluid Dynamics' (2006, Nov 19-21). Moon, R., Vasiliev, V. Tampa, Florida. USA. American Physical Society;
Abstract ID: BAPS.2006.DFD.P19
https://meetings.aps.org/Meeting/DFD06/Content/578
https://meetings.aps.org/Meeting/DFD06/Session/P1.19
http://meetings.aps.org/link/BAPS.2006.DFD.P1.19

[27] The Final Proof of the Michelson Morley Experiment; The explanation of Length Shrinkage and Time Dilation. [Book of materials] 'Scientific Research Center for Ecological Safety of the Russian Academy of Sciences: Scientific Seminar of Ecology and Space'. (2007, February 8-10).
Moon, R. Saint Petersburg, Russia.
https://spcras.ru/en/srcesras/

[28] The Explanation of the Photon's Electric and Magnetic fields and its Particle and Wave Characteristics. [Paper presentation] 'Annual Meeting of the Division of Nuclear Physics Volume 52, Number 10'. (2007, Oct 10-13). Moon, R., Vasiliev, V. Newport News, Virginia. USA. American Physical Society; Abstract ID: BAPS.2007.DNP.BF.15
https://meetings.aps.org/Meeting/DNP07/Session/BF.15
http://meetings.aps.org/Meeting/DNP07
http://meetings.aps.org/link/BAPS.2007.DNP.BF.15

[29] The St. Petersburg State University experiment that discovered the Photon Acceleration Effect. 'Virtual Conference on Nanoscale Science and Technology' VC-NST. (2007, Oct 21-25). Moon, R., Vasiliev, V. University of Arkansas. 222 Physics Building. Fayetteville, AR 72701 USA.
http://www.ibiblio.org/oahost/nst/index.html

[30] The Explanation of Quantum Teleportation and Entanglement Swapping. [Paper presentation] '49th Annual Meeting of the Division of Plasma Physics, Volume 52, Number 11' (2007, Nov 12–16). Moon, R., Vasiliev, V. Orlando, Florida. American Physical Society;
Abstract ID: BAPS.2007.DPP.UP8.21
https://meetings.aps.org/Meeting/DPP07/Content/901
http://meetings.aps.org/link/BAPS.2007.DPP.UP8.21
https://meetings.aps.org/Meeting/DPP07/Session/UP8.21

[31] The Explanation of the Photon's electric and magnetic fields, and its particle and wave characteristics. [Paper presentation]
'60th Annual Meeting of the Division of Fluid Dynamics'. Volume 52, Number 12. (2007, Nov 18–20). Moon, R., Vasiliev, V. Salt Lake City, Utah. American Physical Society;
Abstract ID: BAPS.2007.DFD.JU.22
http://meetings.aps.org/Meeting/DFD07
https://meetings.aps.org/Meeting/DFD07/Session/JU.22
http://meetings.aps.org/link/BAPS.2007.DFD.JU.22

[32] The Explanation of quantum entanglement and entanglement swapping. [Poster Session]
'The 10th International Symposium on the Origin of Matter and the Evolution of the Galaxies (OMEG07) (2007, Dec 4-6) Moon, R., Vasiliev, V. Hokkaido University, Sapporo, Japan. Bibcode: 2008AIPC.1016.....S, Harvard (Astrophysics Data System) ISBN 0735405379
https://ui.adsabs.harvard.edu/abs/2008AIPC.1016.....S/abstract

Books by author {A} and work presented in other published books/booklets

1. *"The Vortex Theory of Matter"* Copyright 1990
   R. Moon. {A} Costa Mesa, California

2. *"The End of The Concept of Time"* Copyright 2000.
   R. Moon. {A} Gordon's Publications of Baton Rouge. Louisiana. ISBN 096792981-4.

3. *"The Bases of the Vortex Theory of Space"* (2002).
   R. Moon. {A} Publishing house; "ZNACK" Director Dr. I. S. Slutskin. Post Office Box 648. Moscow, 101000, Russia. p. 32. (In Russian). Journal ISSN: 2362945.

4. *"The Vortex Theory...The Beginning"* (2003). Copyright 2003.
   R. Moon. {A} (Editor's note by Prof., Dr. Victor V. Vasiliev)
   Gordon's Publications of Fort Lauderdale Fla. USA.

5. *"The Bases of the Vortex Theory"* (2003).
   Book of abstracts: Russian Academy of Sciences; ISBN 5-98340-004-5; TRN: RU0403918096768 OSTI ID: 20530263 R. p. 251. R. Moon. V. Vasiliev
   http://nuclpc1.phys.spbu.ru/nucl/Abstracts/Nucleus_2003.pdf
   http://physics.doi-vt1053.com/ISBN5-98340-004-5/Nucleus_2003.pdf

6. *"The Vortex Theory and some interactions in Nuclear Physics"* (2004).
   Book of abstracts: Russian Academy of Sciences; ISBN 5-9571-0075-7 p. 259.
   R. Moon. V. Vasiliev
   http://nuclpc1.phys.spbu.ru/nucl/Abstracts/Nucleus_2004.pdf
   http://physics.doi-vt1053.com/ISBN5-9571-0075-7/Nucleus_2004.pdf

7. *The Vortex Theory Explains the Quark Theory.* (2005).
   R. Moon. {A} Gordon's Publications of Fort Lauderdale, Florida. USA. p. 205.

8. Dr. Russell Moon PhD Thesis; *"The End of "Time"* Collection of Learned Works Addendum, 2012, (pp. 473-488) VVM Publishing House: ISBN 978-5-9651-0804-6 Editor in Chief: I. S. Ivlev. Saint Petersburg State University. St Petersburg, Russia.
   http://physics.doi-vt1053.com/ISBN978-5-9651-0804-6/Dr-Russell-G-Moon-PhD-thesis-The-End-of-Time.pdf
   http://physics.doi-vt1053.com/ISBN978-5-9651-0804-6/Natural_Anthropogenic_Aerosoles_4pages.pdf

9. *"The Discovery of the Fifth Force in Nature: The Anti-Gravity Force"* Collection of Learned Works (pp. 489-495) R. Moon. M. F. Calvo.
   VVM Publishing House: ISBN 978-5-9651-0804-6 p. 534. Editor in Chief: I. S. Ivlev. St Petersburg State University. St Petersburg, Russia.
   http://physics.doi-vt1053.com/ISBN978-5-9651-0804-6/The-Discovery-of-the-Fifth-Force-in-Nature:-The-Anti-gravity-Force.pdf
   http://physics.doi-vt1053.com/ISBN978-5-9651-0804-6/Natural_Anthropogenic_Aerosoles_4pages.pdf

10. *"The Discovery of the Fifth Force in Nature: The Anti-gravity Force"* Collection of Learned Works (pages 496-503) V. Vasiliev. R. Moon. M. F. Calvo. VVM Publishing House; ISBN 978-5-9651-0804-6 2013. p 534. Editor-in-Chief: I. S. Ivlev, Saint Petersburg State University. St. Petersburg. Russia.
http://physics.doi-vt1053.com/ISBN978-5-9651-0804-6/The-Discovery-of-the-Fifth-Force-in-Nature:-The-Anti-gravity-Force.pdf

Other References

Christopher Scarre. Smithsonian Institution; '*Smithsonian Timelines of the Ancient World*'. p 65. Published September 15th 1993. ISBN-10: 1564583058

Russian Scientific Journals:

1. http://www.new-philosophy.narod.ru/RGM-VVV-RU.htm (in Russian)

2. http://www.new-philosophy.narod.ru/MV-2003.htm (in English)

3. http://www.new-idea.narod.ru/ivte.htm (in English)

4. http://www.new-idea.narod.ru/ivtr.htm (in Russian)

5. http://www.new-philosophy.narod.ru/mm.htm (in Russian)

EPILOGUE
Synchronicities That Lead Russell to the Truth

By Publisher
Stan Clifford

Synchronicity is defined as follows: the simultaneous occurrence of events which appear related but have no discernable causal connection.

In the case of Russell Moon and his discoveries it all could not just be happenstance. When I asked Russell to write a book about himself it was because I knew the magnitude of his discoveries and when the world found out they would all wonder who is this guy? The following is a fairly long list of synchronistic things that happened in his life. If even one of these things did not happen, Russel may not have done what he did. It really makes me wonder if someone or something is pulling strings to make these things happen.

Consider these events so important to Russell Moon's success:

1. An underperforming high school student goes into the navy and gets disciplined.

2. The navy puts Russell in a top-secret mission to repair the most complicated electronic equipment using Boolean algebra. This trains his mind to do some of the most complicated problem solving imaginable.

3. Russell, in an attempt to find pool construction work in Florida, stops on a lonely road at a convenient store and is "forced" by the attendant to read the all-important Reader's Digest article on a near death experience. This changes his life forever.

4. After going back to school, he takes a class that deals with the possibility of higher dimensional space and becomes an expert in understanding how that would work if it does exist.

5. Russell studies every religion and, in doing so, remembers a passage in Luke 17-21 where Jesus refers to an exact description of higher dimensional space as the place where the Kingdom of God is.

6. Russell, struggling to discover what part of the physical world is wrong in science, has a revelational experience where he realizes that it's time – time does not exist as a **key** to the universe but merely as a phenomenon.

7. When no one gives him credit for his book, he has the most synchronistic, random meeting with his future wife, Jeannie, who pushes him to publish the book Russel never knew she had received and read.

8. Russell goes to a science conference of **zany** science concepts and is heard by a Russian physicist who takes Russell's science to the top physicist in Russia – basically discovering him.

9. After being published in Russia, the buzz only stays there for a while until he meets Stan Clifford, a businessman who comes up with a way to communicate Russell's discoveries and make him famous.

10. Finally, after all his discoveries are explained in Chapters 25 and 26, maybe Russell Moon's greatest discovery – the true linking of God and science.

Stan Clifford
CEO
Vortex Publishing, LLC